Jerome Sondericker

Notes on Graphic Statics

With applications to trusses, beams, and arches

Jerome Sondericker

Notes on Graphic Statics
With applications to trusses, beams, and arches

ISBN/EAN: 9783337175382

Printed in Europe, USA, Canada, Australia, Japan

Cover: Foto ©berggeist007 / pixelio.de

More available books at **www.hansebooks.com**

𝕸𝖆𝖘𝖘𝖆𝖈𝖍𝖚𝖘𝖊𝖙𝖙𝖘 𝕴𝖓𝖘𝖙𝖎𝖙𝖚𝖙𝖊 𝖔𝖋 𝕿𝖊𝖈𝖍𝖓𝖔𝖑𝖔𝖌𝖞

NOTES

ON

GRAPHIC STATICS

WITH APPLICATIONS TO

TRUSSES, BEAMS, AND ARCHES

BY

JEROME SONDERICKER

Norwood Press
J. S. CUSHING & CO., PRINTERS
1896

GRAPHIC STATICS.

CHAPTER I.

GENERAL METHODS.

§ 1. *Introduction.*

1. Graphic Statics has for its object the solution of problems in statics by means of geometrical constructions, the results being obtained directly from the scale drawings. The reader is assumed to be familiar with the principles and methods of statics as commonly presented in text-books on Mechanics. These will, however, be briefly stated here, for the case of forces lying in the same plane.

2. Representation of Forces. A force is fully determined when its *magnitude, direction* and *point of application* are known. In dealing with problems in Statics of Rigid Bodies, the *magnitude, direction* and *line of action* of a force are the elements commonly involved, since the equilibrium or motion of such a body is not affected by transferring the point of application of a force to any other point of its line of action.

3. Resultant of Any System of Forces lying in the Same Plane. The *magnitude* and *direction* of the resultant of any system of forces lying in the same plane may be found, I, by representing the given forces by the sides of a polygon taken in order, when the closing side in reverse order is the resultant, or, I$_a$, by resolving each force into components parallel to coördinate axes and combining the resultants of the two sets of components thus formed, the components being treated as if all were applied at the same point.

The *line of action* of the resultant may be found, II, by using the principle that the resultant of two forces lying in the same plane must pass through their point of intersection, or, II$_a$, by the method of moments; the moment of the resultant of any system of forces lying in the same plane being equal to the algebraic sum of the moments of the forces.

1

In case the magnitude of the resultant is zero, the forces will either form a couple or be in equilibrium. If the resultant is a *couple*, its moment can be found by II$_a$, or the given forces can be combined into a single resultant couple by II.

If the lines of action of the given forces intersect at a common point, the line of action of the resultant will pass through this point, its magnitude and direction being found by the methods already stated.

4. Examples. 1. Find the resultant of the four forces F, F', F'', F''' (Fig. 1 A) by each of the preceding methods.

First Solution. Represent the given forces by the sides of the polygon (Fig. 1 B) taken in order; then the closing side, AD, represents the magnitude and direction of the resultant. The numerical values of these quantities may be found, (1) by solving the polygon algebraically, or (2) by direct measurement from a scale drawing.

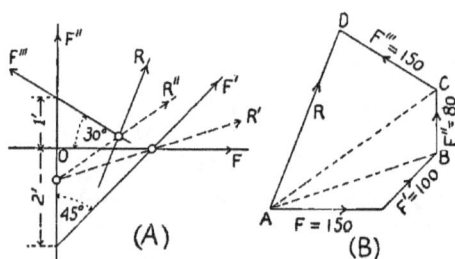

Fig. 1.

To find the line of action of the resultant by the first method, we can proceed as follows: The magnitude and direction of the resultant of F and F'' is AB. Its line of action, R', is drawn parallel to AB through the intersection of the lines of action of F and F'. Continuing in the same manner, we determine R'' to be the line of action of the resultant of F, F', and F'', and finally R to be the line of action of the resultant of the four forces.

Second Solution. Resolve each of the four forces into horizontal (H) and vertical (V) components. Then

$$\Sigma H = 150 + 100 \cos 45° + 0 - 150 \cos 30° = 90.8,$$

$$\Sigma V = 0 + 100 \sin 45° + 80 + 150 \sin 30° = 225.7,$$

$$R = \sqrt{(\Sigma H)^2 + (\Sigma V)^2} = 243.3,$$

$$\alpha_r = \cos^{-1}\left(\frac{\Sigma H}{R}\right) = 68° \ 5'.$$

To find the line of action of R, apply the method of moments. Using O as moment axis, we have

$$\Sigma M = 150 \cdot 0 - 100 \cdot 1.414 + 80 \cdot 0 - 150 \cdot .866 = -271.3.$$

Hence, the moment of the resultant about O is left-handed, and its distance from O is $\dfrac{\Sigma M}{R} = 1.115$. This locates the resultant as given in Fig. 1.

The moment arms of the several forces may be computed or may be measured directly from a scale drawing when the results thus obtained are sufficiently exact.

EXAMPLE 2. Assume four forces, not intersecting at the same point, which form a polygon. Find the resultant couple by each of the two methods mentioned in Art. 3.

5. Conditions of Equilibrium : *Forces not acting at the Same Point.* The *geometrical* conditions of equilibrium for any system of forces lying in the same plane are, I, that the forces can be represented in magnitude and direction by the sides of a polygon taken in order, and, II, that the line of action of the resultant of any portion of the given forces must coincide with the line of action of the resultant of the remainder.

The *algebraic* conditions of equilibrium are, I_a, that if the forces be resolved into components parallel to coördinate axes, the algebraic sum of each set of components must be zero, and, II_a, that the algebraic sum of the moments of the forces must equal zero about any moment axis perpendicular to the plane of the forces.

Conditions I and I_a are equivalent to each other; also II and II_a are equivalent. If I and I_a, only, are satisfied, the resultant is a *couple.*

Forces acting at the Same Point. In this case, condition I or I_a is sufficient. The condition that if three non-parallel forces balance, they must intersect at a common point, is a special case under II.

When a system of forces lying in the same plane is in equilibrium, one or more of the preceding conditions of equilibrium serve to determine the unknown elements of the problem, if it is solvable under the assumption that the body acted on is rigid.

6. Example. The portion of the truss (Fig. 2) to the left of AB is in equilibrium under the action of the supporting force P, load W, and the forces exerted by the portion of the truss to the right of the section AB upon the left-hand portion. The lines of action of these latter forces coincide with the centre lines of the members cut by AB, and their magnitudes are equal to the stresses existing in these members. P and W being known, indicate how to find the three unknown forces by each of the following methods (see Art. 5):

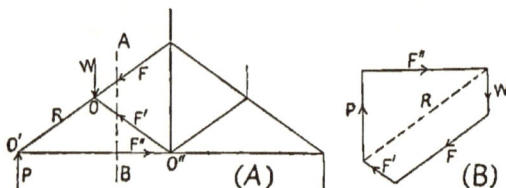

Fig. 2.

(1) By using conditions II and I; (2) by using II_a alone; (3) by using II_a to find one force, then I to find the remaining two; (4) by using II_a to find one force, then I_a to find the remaining two forces.

First Solution. The resultant of P and F''' acting at O' must balance the resultant of W, F, and F'' acting at O. Hence the line of action of each resultant, R, must be OO'. P, F''', and their resultant R, must form a triangle (Fig. 2 B); P being the known force, F''' and R are thus determined. The resultant, R, must balance the forces at O, hence R, W, F, and F'' must form a polygon as shown in Fig. 2 B. F and F', the remaining unknown forces, are thus determined.

Second Solution. To find F, take the moment axis at the intersection O'' of the other two unknown forces, so that their moments will each be zero. Then the algebraic sum of the moments of P, W, and F about O'' must equal zero. Solving the equation thus formed, we determine F. Similarly, to find F'', take moments about O', and to find F''' take moments about O.

Third Solution. We find one force, as F, by the preceding method, then represent the known forces P, W, F, by the sides of a polygon taken in order, and complete the polygon by lines parallel to the two remaining forces F'' and F'''.

Fourth Solution. We find one force F by the method of moments as before. Then, placing the algebraic sums of the horizontal and vertical components of the forces each equal to zero, we form two equations which are solved for the two remaining unknown forces F'' and F'''.

In algebraic solutions the directions of the unknown forces are assumed, the algebraic signs of the results indicating whether the assumed directions are correct or not.

§ 2. *Funicular Polygon*.

7. Definitions. Let F, F', F''' (Fig. 3), be given forces, their magnitudes being represented by AB, BC, CD (Fig. 3 B). Assume any point P, and draw the radial lines PA, PB, etc. From any point M on the line of action of F draw ML and MN parallel to PA and PB respectively. From N, where MN intersects the line of action of F', draw NO parallel to PC; similarly draw OQ parallel to PD, thus forming the broken line $LMNOQ$. We may consider AP and PB, having the directions of the arrows marked 1, to be the components of the force F, the lines of action of these components being ML and MN respectively. Similarly, BP and PC, having the directions marked 2 and the lines of action NM and NO, may be taken as components of F'; and CP and PD with ON and OQ for lines of action as the components of F''. MN is thus the line of action of two equal and opposite forces PB, and NO of the two equal and opposite forces PC. These two pairs of forces consequently balance, leaving AP and PD, having ML and OQ for their lines of action, as the equivalent of the original forces.

The broken line $LMNOQ$ is called a *funicular* or *equilibrium polygon*. The former name is given because the line corresponds to the shape assumed by a weightless cord when fastened at the ends and acted on by the given forces. This is shown by the polygon drawn in dotted lines. The latter name is applicable since a jointed frame of the form of the polygon would be in equilibrium under the action of the given forces.

The point P is called the *pole;* the lines PA, PB, etc., are the *rays;* and the corresponding lines of the funicular polygon are its *strings*.

Figure 3 A is called the *space diagram*, since it represents the location of the lines of action of the forces. Its scale is one of distance. Figure 3 B is the *force diagram*, the lengths of its lines representing the magnitudes of the forces to scale. The perpendicular

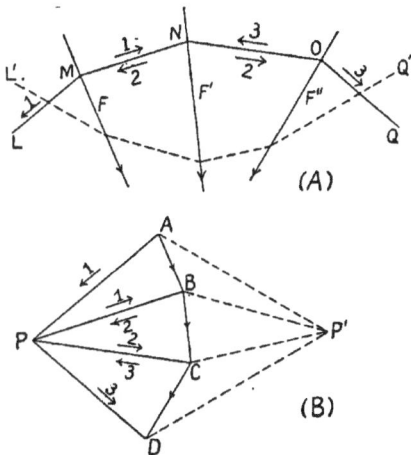

distance from the pole to any side of the force polygon is called the *pole distance* of that force. It is to be noted that this distance represents a force magnitude.

In the case of parallel forces the force polygon becomes a straight line, and the pole distances of all the forces are equal.

8. Applications. The following results are readily derived from the construction explained in the preceding paragraph.

(1) The resultant of F, F', and F'' is given in magnitude and direction by the closing side AD of the force polygon, and its line of action passes through the point of intersection of the strings LM and OQ. The resultant is thus completely determined; and, in general, *the line of action of the resultant of any system of forces, taken consecutively, passes through the point of intersection of the two strings between which the forces lie.*

(2) If the force polygon is closed, PA and PD will coincide, and hence the corresponding strings LM and OQ will be parallel. In this case the resultant is a couple whose arm is the perpendicular distance between the parallel strings, and whose forces are represented by the ray $(PA = PD)$ corresponding to these strings.

(3) In order for the given forces to be in equilibrium, the arm of the couple in (2) must be zero; that is, the strings LM and OQ must coincide in MO. In this case, the funicular polygon is said to be closed. *The conditions of equilibrium*, therefore, *are that both the force and funicular polygons must close.*

If the forces intersect at a common point, the first of these conditions is sufficient, since such a set of forces cannot form a couple.

(4) Any number of funicular polygons may be drawn for the same system of forces by using different poles and beginning the construction of the polygons at different points. Various geometrical relations exist between these different polygons, some of which are given in § 5. The following relations are derived directly from the preceding discussion.

(*a*) Corresponding pairs of non-parallel strings of the various funicular polygons must intersect on the same straight line, this being the line of action of the resultant of the forces included by these strings.

(*b*) If the force polygon is closed, and one funicular polygon closes, all funicular polygons for the given system of forces must close.

In selecting the pole, the obtaining of accurate and convenient diagrams is kept in view. Generally the rays should not make very oblique angles with the adjacent lines of the force polygon. If the pole is taken at a vertex of the force polygon, as at A, Fig. 3 B, the construction of the funicular polygon becomes identical with that explained in Art. 4, for finding the line of action of the resultant by the first method. The construction in Art. 4 frequently leads to inaccurate and inconvenient diagrams and is inapplicable to parallel forces.

9. Notation. To illustrate the notation to be used, let it be required to find the resultant of the four forces ab, bc, cd, de (Fig. 4).

The line of action of any force is represented by the two letters between which it lies; thus, ab represents the line of action of the first force, bc of the second force, etc. In the force diagram, the same letters in capital type are used, but placed at the ends of the lines representing the forces. In constructing the force polygon, the forces and letters must be taken in order from the space diagram (*e.g.* proceeding from left to right). When this is done, if the letters are read in order from left

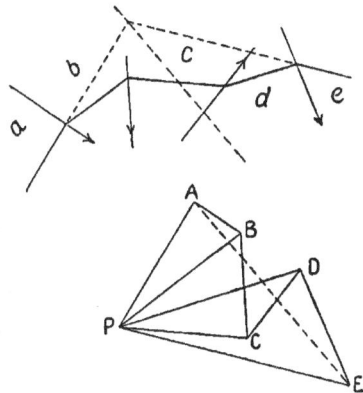

Fig. 4.

to right in the space diagram, the same order of letters in the force polygon indicates the directions in which the successive forces act. It will be noticed that the string corresponding to the ray PA lies in the space a, that corresponding to PB in the space b, etc. These strings will be referred to as the strings a, b, etc.

Having constructed the force and funicular polygons, the magnitude and direction of the resultant is represented by the closing side, AE, of the force polygon, and one point in its line of action is the point of intersection of the similarly lettered strings (a and e). The

Fig. 5.

line of action of the resultant is then drawn through this point of intersection, parallel to AE. In referring to any force (*e.g.* the force BC), the capital letters will be employed, although the line of action of the force as well as its magnitude and direction is included in the reference.

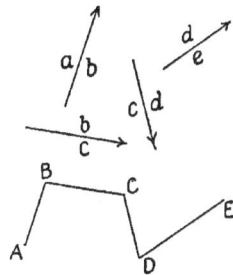

The system of notation just described is applicable in most cases occurring in engineering practice. When the location of the forces is such that it cannot be used to advantage, the modified notation illustrated in Fig. 5 may be employed instead. This needs no description. In general, in referring to forces in the text, the letters will be used so as to indicate by their order the direction in which the forces act.

10. Problems. (1) Assume five non-parallel forces, and find their resultant, using two different poles.

(2) Assume five non-parallel forces whose force polygon closes, and find their resultant, using two different poles.

(3) Assume five parallel forces and find their resultant, using two different poles.

(4) Assume five parallel forces whose algebraic sum is zero and find their resultant, using two different poles.

(5) Assume five parallel forces which balance, and find their resultant, using two different poles.

§ 3. *Forces in Equilibrium.*

11. Use of Funicular Polygon. It will be noticed from § 2 that the funicular polygon construction is especially adapted to the case of parallel and other non-concurrent forces, although it may be used to advantage in problems relating to forces intersecting at a common point, when this point lies outside the limits of the drawing. The conditions of equilibrium, given in Art. 8, for non-concurrent forces are (1) that the force polygon must close, and (2) that the funicular polygon must close. The application of these conditions to the solution of problems will now be illustrated, the cases most frequently arising in practice being presented.

12. Case I. Parallel Forces. Given a system of parallel forces in equilibrium, the lines of action of all and the magnitudes and directions of all but two being known. It is required to find the unknown elements.

Let the unknown forces be the supporting forces of the beam (Fig. 6). Represent the loads in succession by the sides AB, BC, CD of the force polygon, and, selecting a suitable pole, draw the strings a, b, c, d of the funicular polygon. The string a intersects the left reaction at m, and the string d the right reaction at n.

The line joining m, n is then the closing side e of the funicular polygon, and the ray PE drawn parallel to it determines DE and EA to be the right and left hand supporting forces respectively. When the unknown (supporting) forces are parallel, but the given forces (loads) are not, the supporting forces must be parallel to the resultant load.

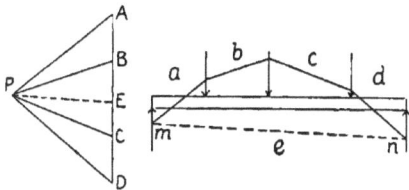

Fig. 6.

If we take AD (Fig. 6) to represent this resultant load, the construction for determining the magnitudes of the supporting forces will be similar to that just explained.

The method of determining the magnitudes and directions of the supporting forces from the lettering, after the construction is completed, should be carefully noted. It is as follows:

(1) The strings are lettered with the same letters as the corresponding rays, so that the strings intersecting on the line of action of any force have the same letters as those which represent that force in the force polygon. The two strings intersecting on the left supporting force are a and e, hence AE represents the magnitude of this force. Similarly d and e intersect on the right supporting force; so its magnitude is DE.

(2) The forces are laid off in the force polygon in right-handed order; *i.e.* if the letters representing them in the space diagram be read in right-handed order, the same order of letters in the force polygon indicates the direction in which the forces act. We also know that the force polygon must close. The force polygon then is $ABCDEA$, the order of letters DE, EA indicating that each supporting force acts upwards. It should be noted in this and the following solutions that *the known forces are made to follow consecutively* in constructing the force polygon.

(To obtain a thorough understanding of these solutions, it would be well for the student to trace out the construction from the standpoint of the triangle of forces. Thus, in the present case, the force AB is resolved into AP and PB, whose lines of action are the strings a and b. Similarly BC is resolved into BP and PC, and CD into CP and PD. The two forces PB and BP balance since they are equal and opposite and have the same line of action, b. PC and CP balance for a similar reason; hence the three loads are equivalent to AP and PD, whose lines of action are a and d respectively. Now, in order to have equilibrium, the resultant of AP and the left supporting force must balance the resultant of *PD* and the right supporting force. The lines of action of these two resultants must then coincide in *mn*. Since, of the three forces intersecting at *m*, one is the resultant of the other two, they must form a triangle, *AP* being the known magnitude. This triangle is *APE*; thus the reaction *EA* is determined. The corresponding triangle for the forces intersecting in *n* is *EPD*.)

13. Case II. Non-parallel Forces. Given a system of forces in equilibrium, all but two of which are known completely. Of these two, the line of action of one and one point in the line of action of the other are known. It is required to determine the unknown elements.

Let the known forces be the wind pressures *AB*, *BC*, *CD* on the roof (Fig. 7). The unknown forces are the reactions of the sup-

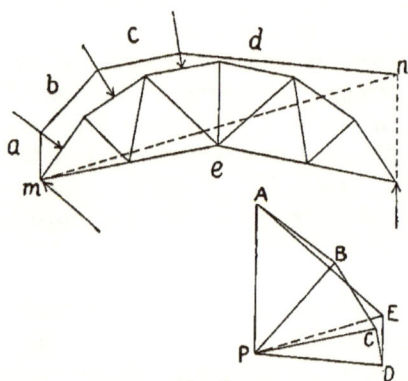

FIG. 7.

ports. The line of action of the right supporting force is given, it being vertical; and one point, *m*, of the left reaction is known. Represent the loads in succession by the sides *AB*, *BC*, *CD* of the force polygon. The strings *e* and *a* must intersect on the line of action of the supporting force *EA*, and as the point *m* of this line of action is known, the funicular polygon will be constructed so that these strings will intersect at this point. The construction of the funicular polygon is then begun by drawing the string *a* through the point *m*, the strings *b*, *c*, *d* being then drawn in order.

The string *d* intersects the right supporting force at *n*, and, of course, the string *a* intersects the left supporting force at *m*; *mn* is thus the closing side, *e*, of the funicular polygon. The ray corresponding to the string *e* is now drawn, intersecting the right supporting force, *DE*, at *E*. The triangle of forces, *PDE*, whose lines of action intersect at *n*, is thus formed, and the magnitude of the supporting force *DE* is determined. Finally, the closing side, *EA*, of the force polygon represents the other supporting force in magnitude and direction. Its line of action is drawn through *m*, parallel to *EA*.

14. Case III. Non-parallel Forces. Given a system of forces in equilibrium, the lines of action of all are known, but the magnitudes and directions of three of the forces are unknown. It is required to determine the unknown elements.

Let ab, bc, cd (Fig. 8) be the lines of action of the three unknown forces. The resultant of any two, as bc and cd, must pass through their point of intersection, n. If this resultant be substituted for bc and cd, the problem becomes identical with

FIG. 8.

Case II; ab being the line of action of one unknown force, and n one point of the line of action of the other.

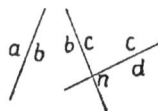

The construction of Case II having been made, the resultant of bc and cd can be resolved into components parallel to these lines of action, thus completing the solution. Figure 9 shows this method applied to determine the stresses in the members ef, fg, and ga, of the truss, the loads and reactions being known.

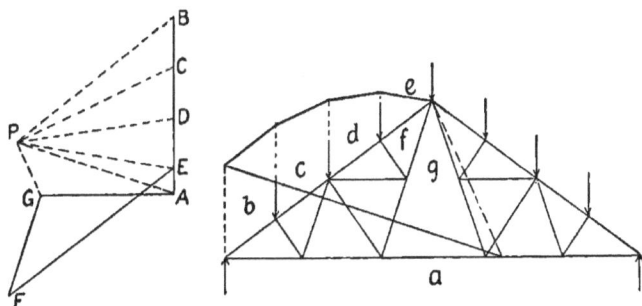

FIG. 9.

15. Resolution of Forces into Components. Since forces equal and opposite to the components and having the same lines of action will balance the given forces, we can solve problems of this nature in the same manner as if the forces balanced, the desired components being the balancing forces with their directions reversed.

§ 4. *Graphical Determination of Moments.*

16. Moment of a Given Force. Let AB (Fig. 10) be any force, and M any moment axis, distant x from the line of action of the force. Assuming any pole, P, resolve AB into the components AP and PB. Draw through M a line parallel to AB, and let PR be perpendicular to AB. The two triangles PAB and OST being similar, we have

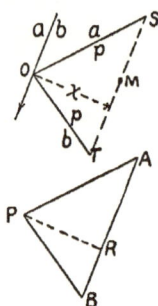

$$AB : PR :: ST : x, \text{ or } AB \cdot x = PR \cdot ST.$$

But $AB \cdot x$ is the moment of the force about M, hence:

The moment of a force about a given axis may be found by drawing a line through the moment axis parallel to the given force, and multiplying the pole distance of the force by the distance intercepted on this line by the two strings corresponding to the force.

FIG. 10.

17. Moment of Resultant of Any System of Forces. The resultant of AB, BC, and CD (Fig. 11) is AD, its line of action passing through the intersection of the strings a and d. The moment of this resultant about any point M is therefore equal to the intercept, ST, times the pole distance, PR, of the resultant. It should be noted that in order to employ this construction the given forces must follow consecutively in the force

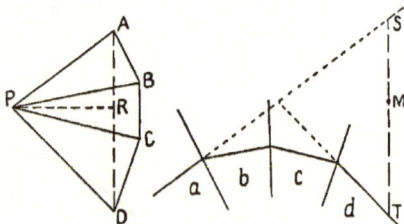

FIG. 11.

polygon. The strings intercepting ST are lettered with the same letters as the resultant force.

18. Moment of Resultant of Parallel Forces. The constructions of this section are especially adapted to the case of parallel forces.

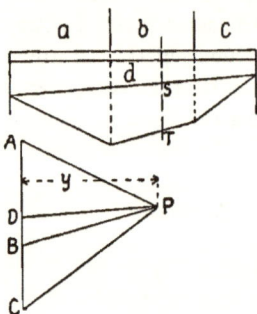

FIG. 12.

Illustration: Let the beam (Fig. 12) be loaded as shown. The reactions of the supports (Art. 12) are found to be CD and DA. It is required to find the bending moment at any section, ST. By definition, this bending moment is equal to the algebraic sum of the moments of the forces to the left of the section, and, by Art. 17, this resultant moment is equal to the pole distance, y, times the intercept ST.

Now in case of parallel forces, the pole distance is the same for all the forces, hence the bending moments at the various sections of the beam are directly proportional to the corresponding intercepts. This subject is further developed in Chapter III.

§ 5. *Geometrical Properties of Funicular Polygons.*

19. Funicular Polygon having the Direction of One String given.
Let it be required to draw a second funicular polygon (Fig. 12) such that the string d will be horizontal. Since this string must be parallel to the ray PD, if we draw a horizontal line through D, a polygon constructed from a pole taken at any point on this line will fulfil the requirement.

20. Intercepts inversely Proportional to Pole Distances. Parallel Forces. It follows directly from Arts. 17 and 18 that if any two funicular polygons be constructed for a given system of parallel forces, the ratio of the intercepts of corresponding pairs of strings on any line drawn parallel to the forces is constant, being equal to the inverse ratio of the pole distances of the respective polygons.

21. Locus of Points of Intersection of Corresponding Strings.
If two funicular polygons be constructed for the same system of forces, their corresponding strings will intersect on a straight line parallel to the line joining the two poles.

Proof. Let AB, BC, CD (Fig. 13) be the given forces, P and P' being the two poles. The four forces AP, PB, BP', $P'A$ balance, since the resultant of AP and PB, *i.e.* AB, is equal and opposite and has the same line of action as the resultant of BP' and $P'A$. Since these forces balance, the resultant of either two, as AP and $P'A$, must balance the resultant of the other two, PB and BP'. The resultant of AP and $P'A$ is $P'P$, and its line of action must pass through the

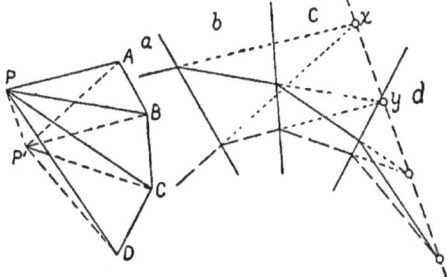

FIG. 13.

intersection x of the corresponding strings. Similarly, the resultant of PB and BP' is PP', and its line of action must pass through y. Since these two resultants balance, their lines of action must coincide in xy, which is therefore parallel to PP'. The intersection of the remaining pairs of strings on xy can be proved in a similar manner.

22. Locus of Poles of Funicular Polygons passing through Two Given Points. It follows directly from the preceding proof that if any other funicular polygon be drawn for the given system of forces such that the string a would pass through x and the string b through y, the corresponding pole would lie on the line PP', and in general : For any given system of forces, the locus of the poles of all funicular polygons, two of whose corresponding strings pass through two given points, is a straight line parallel to the line joining the two points.

23. Funicular Polygon through Three Points. General Solution.

PROBLEM. To draw a funicular polygon for a given system of forces such that three designated strings shall pass through three given points. Let the forces be AB, BC, CD, DE, EF (Fig. 14). It is required to draw a funicular polygon such that the string a will pass through O, the string c through O', and the string f through O''. Let

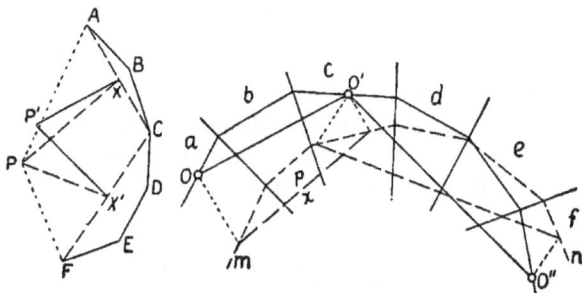

FIG. 14.

mn be any funicular polygon for the given forces, with P for pole. Through O and O' draw lines parallel to AC. These may be taken as the lines of action of two forces which balance AB and BC. To determine their magnitudes draw the closing string px and its corresponding ray PX, giving CX and XA for the required balancing forces. (See Art. 12.) Now, in order for the strings a and c to pass through O and O' respectively, the closing string px must pass through both these points, hence the corresponding pole must lie on the ray $P'X$, drawn through X parallel to OO'. (See Art. 19.) If any point on this line be taken for pole and a funicular polygon drawn such that the string a passes through O, the string c will pass through O'. In a similar manner $P'X'$ is determined as the locus of the poles of all funicular polygons whose strings c and f pass through O' and O'' respectively. The point of intersection P' is then the pole of the required polygon, which is drawn. Evidently a similar construction could be made for the points O and O''. Another method of solving this problem is given in Art. 52.

24. Funicular Polygon through Three Points. Parallel Forces.
A shorter solution than that given in Art. 23 can be made for this

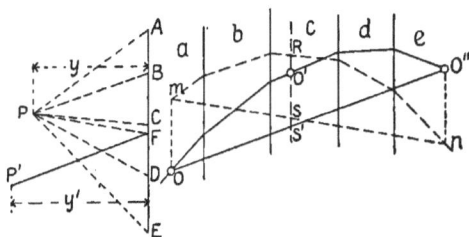

FIG. 15.

case. Let AB, BC, CD, DE (Fig. 15) be the given forces, and O, O', and O'' the given points. Let mn be any funicular polygon for the given forces, P being its pole. Draw lines parallel to the forces through each of the three given points. Assume two of these lines, as the ones through O and O'', to be the lines of action of forces balancing the given forces included between them. The magnitudes of these balancing forces are found by drawing the closing string and parallel to it the ray PF, as in Art. 23. Now, in order for the strings a and e to pass through O and O'' respectively, the closing string must be the line OO'', and the corresponding ray FP' will contain the new pole. The distance intercepted on the line drawn through O' by the first polygon is RS, and by the second polygon will be $O'S'$. The new pole distance, y', can then be found from the proportion $y' : y :: RS : O'S'$ (Art. 20). This locates the pole P' of the required polygon.

Second Method. Instead of using the pole P' in constructing the second polygon, its vertices may be located directly from the fact that the ratio of corresponding intercepts of the two polygons is constant and hence equal to the ratio $RS : O'S'$. (See Art. 20.)

25. Funicular Polygon through Two Points, One String having a Given Direction. PROBLEM.
To draw a funicular polygon for a given system of forces, such that two designated strings shall pass through two given points and one string of the polygon shall have a given direction.

The method of Art. 23 can be used to determine the locus of the poles of all polygons passing through the two given points. The intersection of the ray corresponding to the string whose direction is given with this locus will be the pole of the required polygon.

If the string whose direction is given is also one of the two which are to pass through the given points, the following simpler solution can be made. Let AB, BC, and CD (Fig. 16) be the given forces,

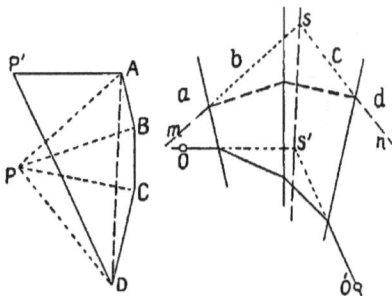

FIG. 16.

and P the pole of any funicular polygon mn for these forces. It is required to draw a second polygon such that the strings a and d will pass through O and O' respectively, and the string a be horizontal. The resultant of the forces included between the designated strings a and d is AD, its line of action passing through the intersection S of the strings a and d of the first polygon. The corresponding strings of the second polygon must intersect on this line (Art. 8). These strings are, therefore, OS', drawn horizontally, and $S'O'$. The new pole P' will be at the intersection of the corresponding rays. The polygon can now be completed.

26. Remarks. In constructing a funicular polygon for a given system of forces, the strings are drawn in succession, the correctness of the location of each depending upon the accuracy with which all the preceding ones have been drawn. In this way errors accumulate,

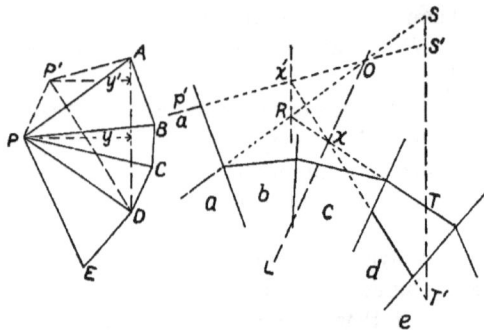

FIG. 17.

and it is necessary to make the construction with extreme care. When, however, one polygon has been drawn, any string of a second polygon can be located directly.

Three special methods of drawing any string of a second polygon are shown in Fig. 17. The pole of the given funicular polygon is P. It is desired to draw any string d of a second polygon, the pole being P' and the first string having the position $p'a$.

First Method. The two strings a intersect at O. Draw OL parallel to PP'. This line will contain the points of intersection of the corresponding strings of the two polygons (Art. 21). The string d' is then drawn through the point of intersection of the string d with OL, and parallel to the ray $P'D$.

Second Method. The strings a and d intersect at R. A line through this point parallel to AD is the line of action of the resultant of the three forces included between these strings. The corresponding strings of the second polygon must intersect on this line. The string a' is known, so d' can at once be drawn through the point of intersection, x', of a' and the line through R.

Third Method. Draw any line ST parallel to the resultant force AD. The strings a and d intercept the distance ST on this line. The moment of the included forces AB, BC, and CD, about any point on this line as moment axis, is then $ST \cdot y$ (Art. 17). This product must evidently be the same for both funicular polygons. The string a' intersects ST at S'. One point of the string d' is then located by laying off $S'T'$, such that $S'T' \cdot y' = ST \cdot y$. A line through T' parallel to the ray $P'D$ is the required string.

The third method is especially adapted to the case of parallel forces. (See Art. 24, Second Method.)

CHAPTER II.

ROOF TRUSSES.

(For construction of roofs, estimation of loads, etc., see Lanza's *Applied Mechanics*, Chap. III.)

§ 1. *Determination of Reactions of Supports.*

27. Direction of Supporting Forces. When a truss is fixed at both ends, the reactions of the supports are indeterminate. They are commonly assumed to be parallel to the resultant load. This assumption, however, is in some cases hardly admissible. (See Art. 28.) When one end is supported on rollers, thus providing for free expansion and contraction, the reaction at the roller end is assumed to be vertical, the direction of the reaction at the fixed end being determined from the conditions of equilibrium.

28. Examples. The methods of determining the supporting forces are included under Arts. 5, 12, and 13. A few examples will be given to indicate suitable solutions.

EXAMPLE 1. The unsymmetrical truss (Fig. 18) is subjected to a dead load uniformly distributed over the upper chord, and snow load on the right side. Find reactions of supports.

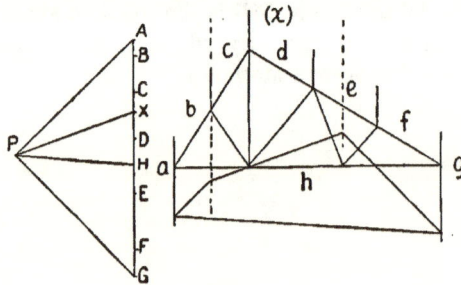

FIG. 18.

The loads supported at the joints are represented in order in the force polygon. The resultant loads on the two sides are AX and

18

XG, their lines of action being given by the dotted lines. The method of Art. 12 is used to determine the reactions, which are *GH* and *HA*. The two resultant loads are used instead of the loads at the joints, so as to reduce the number of sides of the funicular polygon, thus saving labor and increasing the accuracy of the result.

EXAMPLE 2. The truss (Fig. 19) is subjected to wind pressure from the left, the load being uniformly distributed over the rafter. Find reactions of supports: (1) When both ends are fixed (Fig. 19 A); (2) when the right end is supported by a roller (Fig. 19 B); (3) when the left end is supported by a roller (Fig. 19 C).

The method of Art. 12 is used for the first case, and the reactions are also found by dividing the load *AD* into parts inversely proportional to the segments into which the line of action of the resultant load divides the line joining the supports. The method of Art. 13 is used for the second and third cases.

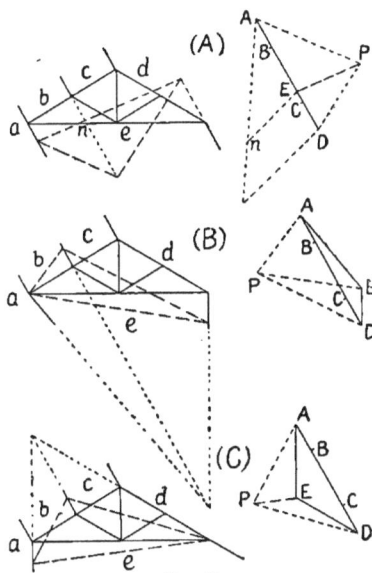

FIG. 19.

In addition, the lines of action of the supporting forces are also determined from the condition that the resultant load and supporting forces (three forces in equilibrium) must intersect at a common point (Art. 5).

EXAMPLE 3. The truss (Fig. 20), fixed at the ends, is loaded with a vertical load at each joint of the upper chord, and also with a wind load on the left side. Find reactions of supports for this combined loading. The resultant wind pressure on segments *b* and *c* acts at their middle points, and this pressure in each case is supported equally at the adjacent joints. At the joint *bc*, for instance, three loads are shown: one-half the pressure on surface *b*, one-half the pressure on surface *c*, and the vertical load supported at the joint.

The loads at the joints are represented in order in the force polygon. The lines of action of the supporting forces are drawn

parallel to the closing side, *AF*, of the polygon. The lines of
action of the resultant loads at the joints are also drawn parallel to
the lines *AB*, *BC*, etc., of the force polygon. These are shown by
dotted lines. A funicular polygon is now drawn for these resultant
joint loads, and the magnitudes, *FG* and *GA*, of the supporting
forces are determined as in the preceding cases.   •

It is to be noted that if the supporting forces (Fig. 20) had been
determined for the dead and wind loads separately, in each case
assuming that the supporting forces are parallel, the resultant

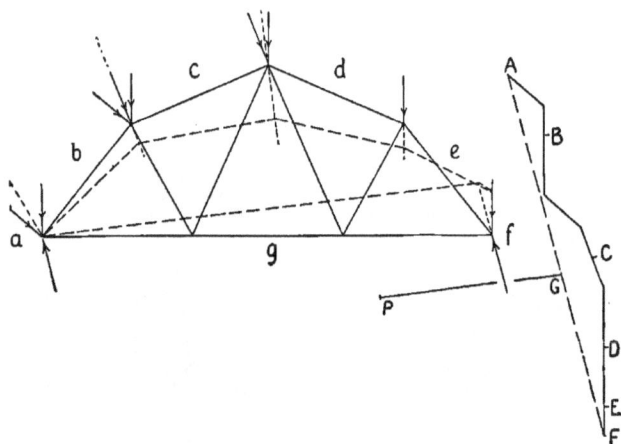

Fig. 20.

reactions would not be parallel, and would therefore differ from
those found in Fig. 20. This difference in results obtained by
dealing with the separate and combined loads does not occur under
any of the modes of treatment discussed other than that of parallel
reactions.

EXAMPLE 4. Another assumption employed in determining the
supporting forces for a truss with fixed ends is that the horizontal
component of the loads is divided equally between the two supports.
The solution of this case is illustrated in Fig. 21. *AB* is the resultant load. The supporting forces are resolved into vertical and
horizontal components, and the horizontal components, *EA* and *BC*,
are assumed equal, and hence each is equal to one-half the horizontal
component of the load. *EA* and *BC* can then be found by drawing a vertical, *CE*, through the middle point of *AB*. *CE* also
represents the sum of the vertical components of the reactions.
These components, *CD* and *DE*, can then be found by dividing *CE*

into parts inversely proportional to the segments of the line joining the supports, or can be found by the funicular polygon. This latter construction is shown in the diagram. The resultant supporting forces are then BD and DA.

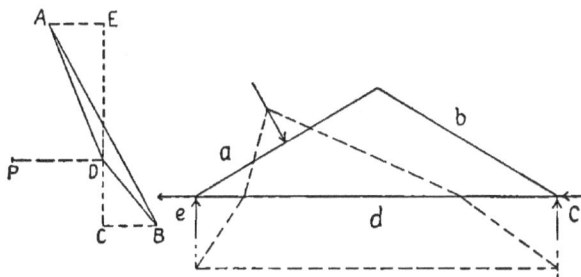

FIG. 21.

Of the two solutions for trusses with fixed ends, illustrated in examples 3 and 4, the latter would appear to be the more generally applicable. For comparatively flat roofs, either solution would seem to be admissible.

EXAMPLE 5. Figure 22 represents a truss supported on columns. The columns are assumed to be hinged at both ends, and the knee-

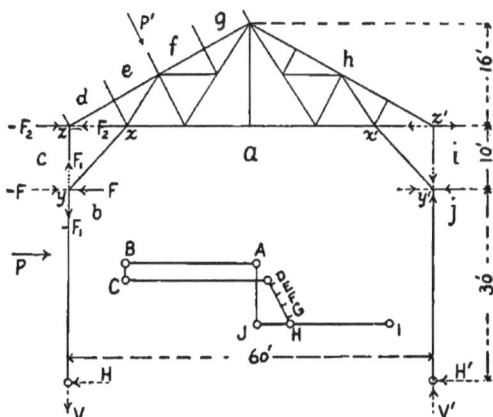

FIG. 22.

braces xy, $x'y'$ prevent distortion under the action of non-vertical forces, such as wind pressure, the force exerted by belts, jib cranes, etc. The sides of the building are assumed to be covered with corrugated iron. This covering is supported by the columns, so that the frame is called upon to resist the pressure of the wind on the side as well as roof of the building. It is required to determine the external forces acting on truss and columns, due to wind pressure.

Distance between trusses	$=16$ ft.
Assumed wind pressure on vertical surface	$=30$ lbs. per sq. ft.
Normal wind pressure on roof (Hutton's formula)	$=19$ lbs. per sq. ft.
Total pressure on side of building (one panel)	$=16 \times 40 \times 30 = 19200$ lbs. $= P$.
Total normal pressure on roof (one panel)	$=16 \times 34 \times 19 = 10336$ lbs. $= P'$.
Horizontal component of P'	$=4864$ lbs.
Vertical component of P'	$=9120$ lbs.

The supporting forces are indeterminate. The assumption made by Hutchinson[1] will usually give a greater stress in the leeward than in the windward knee-brace. A more generally satisfactory treatment would appear to be to assume that the horizontal reactions are divided between the two columns in such a manner as to cause equal stresses in the two knee-braces. The example is solved on this latter assumption.

Taking moments about top of left-hand column, we have

$$\text{Moment of stress in } xy = H \cdot 40 - 19200 \cdot 20.$$

For the right-hand column,

$$\text{Moment of stress in } x'y' = H' \cdot 40.$$

Equating the second members of these two equations, we have

$$H - H' = 9600.$$

From $\Sigma H = 0$ we have

$$H + H' = 19200 + 4864 = 24064.$$

$$\therefore H = 16832 \text{ lbs. and } H' = 7232 \text{ lbs.}$$

The vertical reactions can now be found by moments. Taking the foot of the right-hand column as moment axis, we have

$$V \cdot 60 = 19200 \cdot 20 + 4864 \cdot 48 - 9120 \cdot 45.$$

$$\therefore V = 3451 \text{ lbs. and } V' = 3451 + 9120 = 12571 \text{ lbs.}$$

The horizontal forces at y and z which balance H and P are found by moments to be $F = 28928$ lbs. and $F_2 = 31296$ lbs. We can now determine the forces which act upon the column, and those which act on the truss, by applying two equal and opposite forces, $F = 28928$ lbs. at y, and two equal and opposite forces, $F_2 = 31296$ lbs. at z. Also apply at y two equal and opposite vertical forces, $F_1 = V = 3451$ lbs. These added forces will not affect the stresses.

---

[1] See Johnson's *Modern Framed Structures*, Chap. XXIX., or *Proceedings Engineers' Society of Western Pennsylvania*, Vol. VIII., p. 247.

We now have acting on the column a set of balanced forces, H, P, $-F$, and F_2, causing a bending stress; also the balanced forces V and F_1 causing tension in the lower portion of the column. The forces acting on the truss, taking y and y' as the points of support, are: at y, F and $-F_1$; and at z, $-F_2$. The right-hand column is dealt with similarly, the dotted arrows giving the forces acting on the column, and those drawn in full lines, the forces acting on the truss. The polygon of external forces for the truss is plotted to scale in Fig. 22, $HI = 21696$ lbs., $IJ = 28928$ lbs.

If the columns are rigidly fixed at the base, the fastening will assist the knee-braces in resisting distortion, and the stresses in knee-braces and truss will be reduced.[1]

When a truss is loaded at both upper and lower chords, as in Fig. 1, Plate I, a difficulty occurs in using the funicular polygon construction to determine the supporting forces. In representing the external forces in the force polygon, they are laid off in succession, proceeding around the truss from left to right, i.e. in the order $AB \cdots IJK \cdots RA$. This is done so as to be able to determine the stresses conveniently. On the other hand, in determining the supporting forces by the usual funicular polygon construction, the known forces must follow consecutively. This would require the construction of a second force polygon. No difficulty will be experienced in finding the reactions by other methods in such cases.

29. Algebraic Methods. Frequently the supporting forces can be computed readily, using the algebraic methods of Art. 5 or solving the geometrical figures of the graphical constructions. Such computations should be employed freely in connection with graphical solutions, for the purpose of securing a greater degree of accuracy, and as checks on the constructions.

PROBLEM 1. Indicate in detail a suitable method for computing the reactions of the supports (Fig. 18).

PROBLEM 2. Do the same for Fig. 20.

§ 2. *Determination of Stresses.*

30. General Methods. A truss is designed to support loads applied at the joints, by virtue of the resistance to extension and compression of its various members.

There are two general methods for determining the tension and

---

[1] See Johnson's *Modern Framed Structures*, p. 159.

compression stresses in the members of a truss: (1) method of sections; (2) method of joints.

31. Method of Sections. Let the imaginary line xy (Fig. 23) divide the truss into two parts, this line intersecting the three members CI, IH, HG. The portion of the truss to the left of xy is a body in equilibrium under the action of certain forces. Consider the member CI. If it is in tension, the portion to the right of xy must be exerting force upon the portion to the left of xy, toward the right.

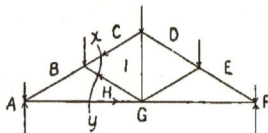

FIG. 23.

If CI is in compression, its right-hand portion must, on the other hand, be exerting force upon its left-hand portion, towards the left. In either case, the magnitude of the force is equal to that of the stress in the member, and the line of action of the force has the direction of the length of the member. This force is external with reference to the portion of the truss to the left of xy. It will be represented by the letters CI. Similar explanations hold for IH and HG. The forces, therefore, which hold the left portion of the truss in equilibrium are the known forces GA, AB, BC, and the forces, CI, IH, and HG exerted by the right-hand portion of these three members upon their left-hand portion. The lines of action of these last three forces are known, their magnitudes and directions being unknown. These six forces constitute a system of forces in equilibrium, lying in the same plane but not acting at the same point. One or more of the conditions of equilibrium of Art. 5 can therefore be used to determine the unknown forces, as was indicated in Art. 6. Of these solutions, the method of moments is generally most useful in dealing with roof trusses.

The following points should be noted:

(1) Only three forces, unknown in magnitude, can be determined, so that if the section xy cuts more than three members which are in action under the given loads, a solution cannot be made.

(2) The stress in a member is equal to the magnitude of the force which it exerts, and the nature of the stress, tension, or compression can be determined from the direction of the force (tension, if directed away from left portion, otherwise compression).

32. Method of Joints. The external forces acting upon the joint A (Fig. 23) are: (1) the supporting force GA; (2) the load AB; (3) the forces exerted by the members BH and HG upon the joint. If

either member, as *BH*, is in tension, the force which it exerts on the
joint *A* is evidently directed away from the joint; if in compression,
towards it. As the joint is in equilibrium these four forces must
balance, and by applying either the algebraic method of resolution
of forces or the geometric method of polygon of forces, the unknown
forces can be determined.

In using the method of joints, the following points should be
noted:

(1) The forces dealt with are those acting on the joint.

(2) In dealing with any one joint only two unknown forces can
be determined.

(3) The nature of the stress in a member can be determined from
the direction of the force which the member exerts on the joint; tension, if the force acts away from the joint; compression, if the force
acts towards the joint.

(4) When the stress in a member is determined, the force which
it exerts upon the joint at each end is known, these forces being
equal and opposite.

33. Determination of Stresses in Roof Trusses. Of the methods
which have been explained, the one best adapted to roof trusses is
the method of joints, solving by the polygon of forces. In applying
this method, the external forces being known, we begin by constructing the polygon of forces at any joint of the truss where only two
stresses are unknown. Having thus determined these two, we repeat
the construction for another joint where only two stresses remain
unknown, and continue in this manner until the stresses in all the
members have been determined.

34. Example. Bow's Notation. The truss (Fig. 24) is subjected
to a uniformly distributed load, *W*. Each intermediate joint of the
upper chord supports $\frac{1}{4} W$, and each end
joint, $\frac{1}{8} W$. Each supporting force is $\frac{1}{2} W$.
The external forces are lettered as in the
preceding chapter, *ab* representing the
left reaction; *bc*, *cd*, etc., the successive
loads. Letters are also placed in the
spaces into which the surface is divided
by the web numbers. Each member of
the truss is represented by the letters in
the adjacent spaces; *e.g.* the two halves
of the lower chord are *ka* and *ha*; the
vertical member is *ji*, etc. The direc-

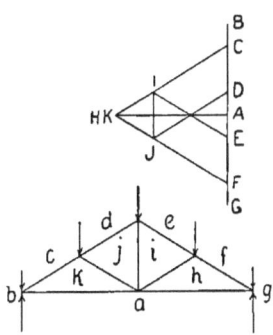

FIG. 24.

tions of the forces are indicated in the force or stress diagram by the order of the letters. For this purpose the letters are to be read in *right-handed* order around the truss or any joint of it; *e.g.* the left reaction is *ab*, left-hand load *bc*, etc.; the force exerted by the member *ck* upon the left-hand joint is *ck*, and upon the joint *cdjk* is *kc*.

To illustrate the manner of using this notation, the stresses in the various members of the truss will be determined. First construct the polygon of external forces. Lay off the loads on a vertical line in right-handed order. The first load, beginning at the left, is *bc*; it acts downwards, and to indicate this, the force is lettered so that *BC* reads downwards. Continuing in this manner, when the polygon is completed, if we read the letters surrounding the truss in right-handed order, *bcdefgab*, these letters in the stress diagram, read in the same order, will represent the polygon of external forces, the order of succession of the letters indicating the direction in which the forces act.

Next the polygon of forces for the left-hand joint of the truss is constructed. The known forces *ab* and *bc* are already included in the stress diagram. The two unknown forces are *ck* and *ka*. One letter of each (*C* and *A*) is already in the stress diagram. These letters indicate the points from which to draw the lines representing the unknown forces. Thus the polygon is completed by drawing from *C* and *A* lines parallel to *ck* and *ka* respectively, their point of intersection being lettered *K*. Proceeding to the next joint, *kc* and *cd* are the known forces. From the points *D* and *K*, lines are drawn parallel to *dj* and *jk* respectively, intersecting at *J*, thus completing that polygon. By this mode of construction, the polygons of forces for the various joints of the truss are grouped together in a single diagram called the *stress diagram*. When the last joint is reached, all the stresses but one will be known. The letters representing this one will already be in the stress diagram, and the line joining them will be parallel to the corresponding member of the truss if the construction is accurate.

The following points should be noted in using Bow's notation :

(1) The stress in any member of the truss is represented by the same letters as the member itself.

(2) The polygon of forces for any joint is lettered with the letters surrounding that joint, the direction of the forces being indicated by the succession of letters obtained by reading the letters surrounding the joint in right-handed order.

(3) To determine whether any member is in tension or compression, read the letters representing that member in right-handed order

about the joint at either end of the member. The same order of succession of these letters in the stress diagram indicates the direction of the force which the member exerts on the joint used. If this direction is towards the joint, the stress is compression; if away from it, tension. As an illustration, let the nature of the stress in *ck* be determined. Using the left joint, the letters read *c-k*. In the stress diagram, this order of succession of these letters is towards the left. Referring again to the truss, the direction, towards the left, is seen to be towards the joint used as a centre, this indicating that the member *ck* is exerting force on this joint towards it. Hence *ck* is in compression.

(4) In constructing the polygons of forces it is to be noted that all the sides of any polygon but the two corresponding to the two unknown forces at the joint will be already represented in the stress diagram, and the points from which to draw these two sides are indicated by the lettering, as previously explained. The point of intersection of these two sides is marked with the letter common to the corresponding members of the truss.

In using the method illustrated, the truss is drawn accurately to scale, the loads are also represented accurately to scale, and the stress diagram is constructed with extreme care. The magnitudes of the stresses are gotten by scaling off the proper lines of the stress diagram, and the nature of the stress is derived from the lettering, as previously explained. The most serious cause of inaccuracy, especially in case of trusses having a large number of members, is that the construction of each polygon in turn is based upon preceding ones, so that errors accumulate. To guard against this, it is well to determine the stress in one or more of the members analytically. The method of moments is especially adapted to such uses. (For detailed directions regarding the work of constructing the diagrams, see § 7.)

35. Example 1. Wooden Truss fixed at Both Ends. Figure 1, Plate I, represents one of a series of parallel trusses, spaced 16 feet between centres, supporting a roof. The vertical iron tie-rods divide it into 8 panels of equal width. Span = 80 ft. Rise of upper chord = 16 ft. The roof is to be covered with tin laid on sheathing 1 in. thick. The sheathing is supported by rafters 2 in. by 7 in. section, spaced 2 ft. between centres. The rafters are supported by purlins 8 in. by 12 in. section, these being supported at the joints of the upper chord.

Each rafter supports an area of the roof surface 2 ft. wide by

10.8 ft. long, and is proportioned as a beam to support the maximum load, including its own weight, which is liable to come upon this area. Each purlin supports an area 10.8 ft. wide by 16 ft. long, and is also proportioned as a beam for the maximum load to which it is liable to be subjected. The student should verify these dimensions of rafters and purlins, using a working stress of 1000 lbs. per sq. inch.

The snow load is assumed to be 20 lbs. per sq. ft. of horizontal projection of roof surface, and the wind pressure 40 lbs. per sq. ft. on a vertical surface.

CALCULATION OF LOADS.

Wt. of tin and sheathing per sq. ft. of roof surface $= 3\frac{1}{2}$ lbs.
"   " rafters (30 lbs. per cu. ft.)   " $= 1\frac{1}{2}$ "
"   " purlins (30 lbs. per cu. ft.)   " $= 2$ "

Total $= 7$ lbs.

Estimated weight of truss, $W = \frac{1}{2} al (1 + \frac{1}{10} l) = 5760$ lbs. This weight is assumed to be supported at the joints of the upper chord.

Normal component of wind pressure (Hutton's formula) $= 20$ lbs. per sq. ft. of roof surface.

(The weight of the truss must be estimated from the actual weights of trusses already built. Merriman gives the formula for wooden trusses, $W = \frac{1}{2} al (1 + \frac{1}{10} l)$, and for iron trusses, $W = \frac{3}{4} al (1 + \frac{1}{10} l)$. These formulas are derived from a table of weights of trusses given by Ricker. Johnson gives, for iron trusses, the formula, $W = \frac{1}{24} al^2$. In each case $W =$ weight of truss in pounds, $a =$ distance between trusses in feet, $l =$ span in feet. These formulas can, at the best, be only rough approximations.)

LOADS AT INTERMEDIATE JOINTS OF UPPER CHORD.

Roof covering, etc. $7 \times 16 \times 10.8 = 1210$ lbs.
Truss $\frac{1}{8} \times 5760 = 720$ "

Total dead load at each intermediate joint $= 1930$ lbs.
Snow   "   "   "   "   " $= 20 \times 16 \times 10 = 3200$ "
Wind   "   "   "   "   " $= 20 \times 16 \times 10.8 = 3460$ "

Each extreme joint supports one-half the load carried by an intermediate joint.

The roof is also to support a ceiling which is suspended from the joints of the *lower* chord. Its weight, estimated in the same manner as the preceding, is 17 lbs. per sq. ft., or 2720 lbs. per joint.

36. Stresses. The general method of determining the stresses has already been explained. The student should trace out the construction of the diagrams.

Dead Load Stress Diagram. The reaction of each support is $\frac{1}{2}(1930 \times 8 + 2720 \times 7) = 17240$ lbs. The loads and reactions are laid off in right-handed order, and the polygons of forces for the various joints are constructed, beginning at the left support. Since the truss is symmetrical and symmetrically loaded, the stresses in corresponding members of the two halves will be equal, so the stresses for one side only are determined. The following special precautions were taken to ensure accuracy:

(1) The intersection of BS and RS is quite oblique, and since the accuracy of the diagram depends upon the correctness of location of the point S, it is well to determine that point by computation as follows: The forces at the left support form a triangle similar to that formed by the half truss. Hence,

$$RA - AB\,(16275 \text{ lbs.}) : SR :: 16 : 40. \quad \therefore SR = 40687 \text{ lbs.}$$

This serves to locate S accurately. The remaining stresses are then determined graphically.

(2) In order to check the completed diagram, the stress in zf is computed by moments. (See Art. 6, Second Solution.) Taking the middle lower joint as moment axis and dealing with the left-hand forces, we have

$$FZ \times 14.856 = 16275 \times 40 - 4650 \times 30 - 4650 \times 20 - 4650 \times 10.$$

$$\therefore FZ = 25040 \text{ lbs.}$$

The value of this stress as measured from the diagram is 25000 lbs.

Snow Stress Diagram. This diagram needs no explanation. The same precautions were taken as in the preceding case. The computed stress in zf is 17232 lbs., the diagram scaling 17200 lbs.

Wind Stress Diagram. This is drawn for wind on the left side. The total wind pressure is $4 \times 3460 = 13840$ lbs. As both ends are fixed, the reactions are assumed to be parallel to the resultant wind pressure, and their magnitudes, computed by the principles of parallel forces, are 9818 lbs. and 4022 lbs. The force polygons are constructed, working from each end towards the centre of the truss. It is to be noted that the stresses in all the web members on the lee side are zero. The diagram checks itself by the closing of the last polygon of forces. All complete diagrams afford such a check. In addition, the stress in $n1$ was found by moments to be 7827 lbs., the diagram giving 7800 lbs.

37. Maximum Stresses. The members of a truss are proportioned to withstand safely the maximum stresses to which they are liable to be subjected. Having determined the stresses resulting from the dead load, wind, snow, etc., the different combinations of these loads liable to occur must be decided upon, and the corresponding stresses found by taking the algebraic sum of the stresses due to the separate loadings. The maximum result is then the stress to be used in estimating the dimensions of the members. The algebraic minimum stress should also be noted. If the stress in a member changes sign, the member must be designed to resist both tension and compression, also both extreme stresses are used when the working stress is determined in accordance with the laws of failure under repetition of stress.

If it be assumed that maximum wind and snow loads cannot occur at the same time, we would have the following systems of loads to consider : (1) dead load; (2) dead and snow loads; (3) dead and wind loads. If the truss is unsymmetrical, or supported differently at the two ends, the dead load must be combined with wind load on each side. In the design of the arches supporting the train-shed •roof of the Philadelphia and Reading Terminal Railway, Philadelphia, the stresses were determined for: (1) dead load; (2) dead load and snow on one side; (3) dead load and snow on both sides; (4) dead load and wind on one side; (5) Dead and snow loads with wind on one side.

38. Tabulation of Stresses. The stresses are scaled off from the diagrams and inserted in the table, page 31. The sign of the stress is determined from the lettering. (See Art. 34.) The combinations of stresses considered are : (1) dead load; (2) dead and snow loads; (3) dead load and wind on either side; (4) dead and snow loads with wind on either side. The maximum and minimum results are recorded in the table. In this example, the maximum stresses for corresponding members on the two sides of the truss are evidently equal.

For convenience of reference the stresses may also be given as shown in Fig. 1 d, Plate I. The double lines indicate compression numbers.

39. Example 2. Fig. 2, Plate I, is an iron truss, one end resting on rollers. (In practice rollers would not be used for so short a span.)

TABLE OF STRESSES.

Compression +; Tension −.

Member.	Dead Load.	Snow.	Wind.	Max.	Min.
$i6 = bs$	+ 43820	+ 30200	+ 20170	+ 94190	+ 43820
$h4 = cu$	+ 37600	+ 25850	+ 16550	+ 80000	+ 37600
$g2 = dw$	+ 31300	+ 21570	+ 12900	+ 65770	+ 31300
$fz = ey$	+ 25050	+ 17250	+ 10000	+ 52300	+ 25050
$6k = rs$	− 40700	− 28000	− 21770	− 90470	− 40700
$5l = qt$	− 40700	− 28000	− 21770	− 90470	− 40700
$3m = pv$	− 34900	− 24000	− 17100	− 76000	− 34900
$1n = ox$	− 29050	− 20000	− 12450	− 61500	− 29050
$56 = st$	− 2720	0	0	− 2720	− 2720
$34 = uv$	− 5050	− 1600	− 1850	− 8500	− 5050
$12 = wx$	− 7370	− 3200	− 3720	− 14290	− 7370
yz	− 16700	− 9600	− 5570	− 31870	− 16700
$45 = tu$	+ 6220	+ 4320	+ 5000	+ 15540	+ 6220
$23 = vw$	+ 7450	+ 5100	+ 5970	+ 18520	+ 7450
$z1 = xy$	+ 9050	+ 6220	+ 7270	+ 22540	+ 9050

Span = 50 ft. Rise of lower chord = 2 ft. Rise of upper chord = 12 ft. Distance between trusses = 16 ft. Length of middle panel of upper chord = 12 ft. The divisions of upper and lower chords are otherwise equal. The ventilator roof is raised 4 ft.

Dead load (total) = 10000 lbs. = 2000 lbs. per panel of upper chord.
Snow load (15 lbs. per sq. ft. hor. proj.) = 2400 lbs. per panel of upper chord.
Wind load = 40 lbs. per sq. ft. on vertical surface = 31 lbs. per sq. ft. on slope.
Total wind load on main slope of roof = 11150 lbs.
Total wind load on slope of ventilator roof = 3520 lbs.
Total wind load on vertical surface = 2560 lbs.

Attention is directed to the following points:

(1) The ventilator roof, braced as shown by the dotted lines, forms a complete truss fixed at the ends. The pressures on the two

supporting joints of the main truss, due to loads on the raised portion, are then determined in the same manner as are the reactions of a truss with fixed ends. Thus the resultant wind pressure on the raised portion, for wind on the left side, is $a-F$, its line of action cutting el at x. The resulting pressures, aE and EF, on the two joints of the main truss are found by dividing aF into parts inversely proportional to the segments of el. The wind pressure on the joint de is then the resultant of aE and Da, the latter being one-half the wind pressure on the panel d.

(2) In finding the supporting forces due to wind pressure by the funicular polygon, the resultant of the pressures on the main roof, and on the vertical surface of the raised portion, was used in order to obtain better intersections than otherwise.

(3) The snow load is one-fifth greater than the dead load. Since both are assumed to be distributed in the same manner, the snow load stresses can be found by multiplying the corresponding dead load stresses by six-fifths, no snow load diagram being necessary.

(4) The maximum and minimum stresses are found in a manner similar to those of Example 1, with the exception that both sets of wind stresses must be considered. The results are given in Fig. 2 d. The stresses in kl and lm change sign. It will be seen from the diagrams that wind on the fixed side causes compression in kl and tension in lm, while the reverse stresses result from wind on the other side.

(5) The stresses in the members of the ventilator roof truss can be found by treating it as a truss with fixed ends.

The student should trace out the construction of the diagrams and verify the maximum and minimum stresses.

40. Trusses having only Two Unknown Forces to determine at Each Joint. In the preceding examples only two unknown stresses were encountered at each joint, so that the polygons of forces for the successive joints of the truss could be constructed at once. Other cases of the same nature are given in Fig. 25 A, B, C, D, E, F, G, and H.

41. Fink or French Truss. (See Fig. 26.) This is a form of iron roof truss in common use for shops and similar buildings. In constructing the force polygons for this truss, a difficulty is encountered to which attention is directed. Beginning at either (the right-hand) support, the force polygons can be constructed in the usual manner until the joint $ruvk$ is reached, when three unknown forces are

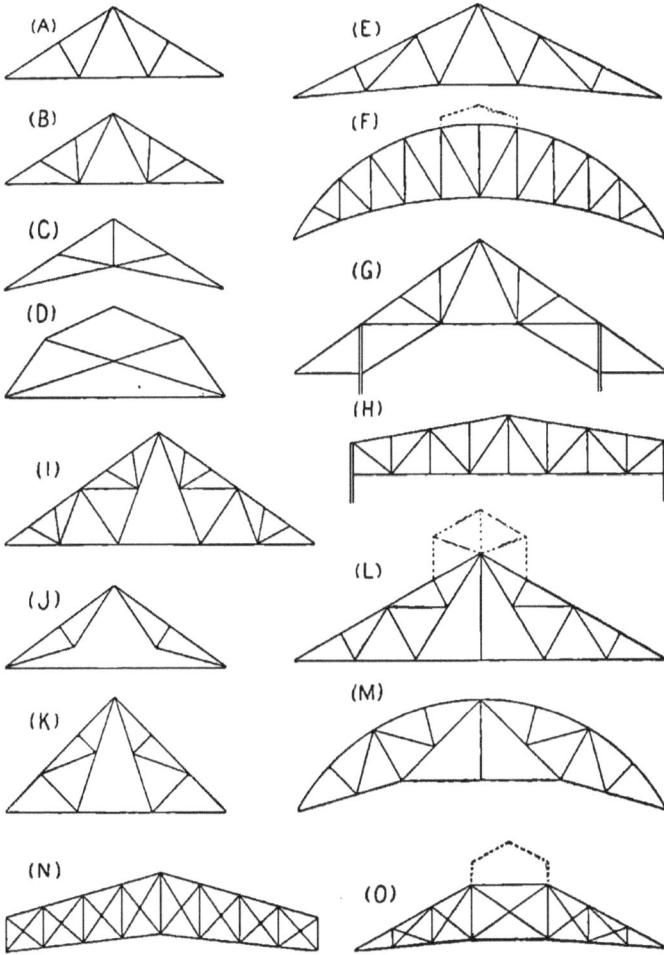

Fig. 25.

encountered. To overcome this difficulty, the stress in rk must be found by some other method.

The most satisfactory solution is to calculate the stress in rk by moments, as explained in Art. 6, and insert this calculated stress in

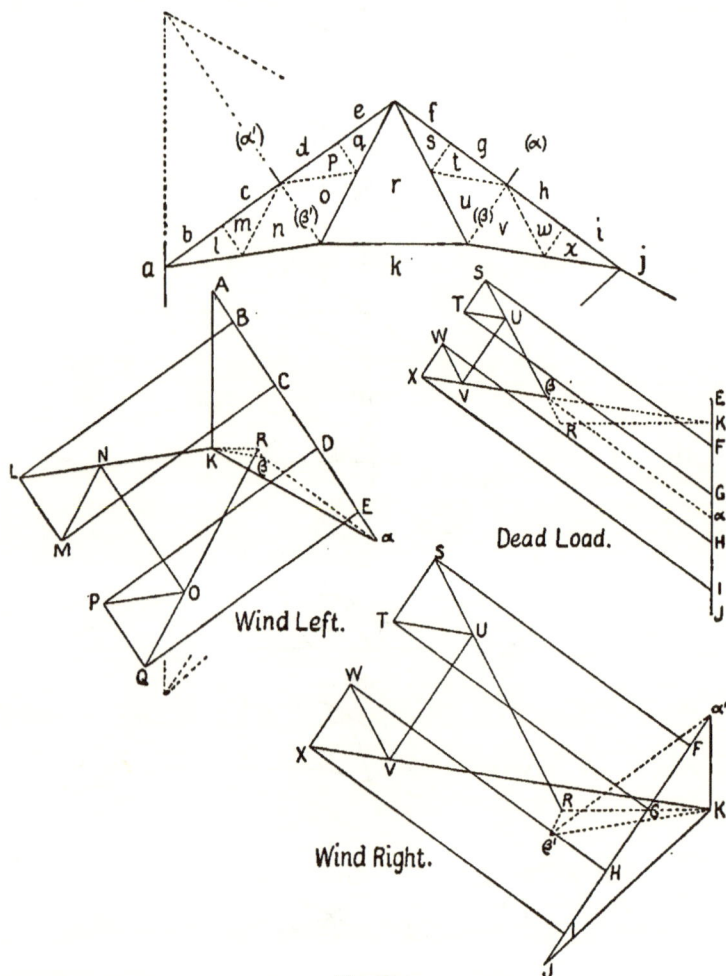

Fig. 26.

the stress diagram. Then the stresses in the two remaining members, ru and uv, can be found by completing the force polygon.

A general graphical solution is shown in Fig. 9. (See Art. 14.)

A simpler graphical solution, shown by dotted lines in the stress diagrams of Fig. 26, is as follows: Since, from the solution by moments, the stress in rk depends only upon the moments of the external forces acting on either half of the truss, about the apex of the truss as moment axis, a single load acting at the end joint can be substituted for the given loading on that side provided that the moment about the apex is not altered. When the load is uniformly distributed over the rafter, this single load will evidently be equal to one-half the load on the rafter. Under this altered loading the stresses in the dotted members will be zero, since they serve only to support loads at the intermediate joints of the rafters. The rafter may then be treated as a single member, $\beta\alpha$. Referring to the dead load stresses, the forces acting at the right end joint under these conditions are $\beta\alpha$, $\alpha j = \frac{1}{4}$ W, jk, and $k\beta$. The forces $k\beta$ and $\beta\alpha$ are then determined by completing the force polygon $\alpha JK\beta\alpha$. Next, the force polygon for the joint βkr is completed, thus determining the true stress KR, and locating R in the stress diagram.

In the wind diagrams this construction is applicable to the unloaded side without altering the distribution of the loads, since the dotted members on that side are not stressed, for the reason stated above.

It is to be noted that this method is only applicable when the two halves of the upper chord are straight lines, hence it could not be employed for such a truss as Fig. 25 M.

Other forms of trusses requiring the same special treatment as Fig. 26 are shown in Fig. 25 I, J, K, L, and M.

§ 3. Counterbracing.

42. Definitions. It will be noticed that the members of a complete truss form a system of triangles. The triangle is the elementary truss. Under the action of forces lying in its plane and acting at its vertices, it cannot be distorted without changing the lengths of one or more of its sides, and such changes are opposed by the resistance to extension and compression of the sides. A polygon of more than three sides, assumed free to turn at the joints, can be distorted without altering the lengths of any of the sides.

The rectangular frame (Fig. 27), acted on by the force F, would be distorted as shown, the distance AB becoming shorter and CD longer, change in length of the diagonals necessarily accompanying the distortion of the frame. A diagonal member capable of resisting both extension and compression would, then, prevent dis-

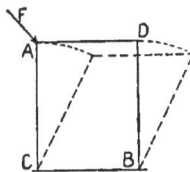

FIG. 27.

tortion. A member joining *C* and *D*, capable of resisting tension alone, would prevent distortion in the direction shown in Fig. 27, but not in the opposite direction. Two tension diagonals, however, would evidently make the frame stable. Thus a rectangular frame may be made capable of resisting any forces acting at the angles, tending to distort it, by the introduction of a single diagonal member capable of resisting both tension and compression, or of two diagonals, both capable of resisting tension alone or compression alone. In the case of two diagonals it is evident that only one of them would be stressed at a time. Considering the quadrilateral of Fig. 27 to represent one panel of a truss, the diagonal which is stressed under the action of the dead load is called the *main brace;* the other, the *counterbrace* or *counter.* The counter may be stressed under the action of the wind, snow on one side of the roof, or other non-symmetrical temporary load.

43. Determination of Stresses. The work of determining the stresses in trusses having counterbracing is more involved than is the case with the class of trusses previously considered. The basis of the treatment is the fact that only one diagonal of a panel is stressed at a time. Two methods of solution will be illustrated in the following example, each possessing certain advantages.

EXAMPLE. The chords of the Crescent truss (Plate II) are circular arcs. The left end is on rollers. Span = 80 ft. Rise of upper chord = 26 ft. Rise of lower chord = 10 ft. Distance between trusses = 20 ft. The truss is divided by verticals into 5 panels, each 16 ft. wide. The diagonals are to be tension members. Assumed weight of truss = 10800 lbs. Weight of roof surface, etc. = 20 lbs. per sq. ft. Snow load = 20 lbs. per sq. ft. of horizontal projection. Wind load = 40 lbs. per sq. ft. on a vertical surface. The wind pressure on each panel is assumed to act at right angles to the chord of the arc. The joint loads used are given in the following table.

TABLE OF JOINT LOADS.

JOINT.	TRUSS.	ROOF, ETC.	TOTAL DEAD LOAD.	SNOW.	WIND R.	WIND L.
AB	lbs. 1080	4940	6020	3200		9390
BC	2160	8390	10550	6400		{ 9390 { 3450
CD	2160	6650	8810	6400		3450
DE	2160	6650	8810	6400	3450	
EF	2160	8390	10550	6400	{ 3450 { 9390	
FG	1080	4940	6020	3200	9390	

44. Lettering Truss Diagram. An adaptation of Bow's notation is used. One diagonal of each panel is dotted. The usual method of lettering is followed, with the exception of the panels having counterbracing, *e.g.* *mn* represents the diagonal of the second panel, which is drawn in full lines, while the same letters accented will be used to represent the dotted diagonal. While we are dealing with either diagonal of any panel, the other is considered to be removed, since only one of them is stressed at a time. If we are dealing with *mn* and *l'k'*, for example, the vertical will be represented by *nl'*; the segments of the upper chord will be *cn* and *dl'*; the segments of the lower chord, *mr* and *k'r*. Again, when dealing with the diagonals *m'n'* and *l'k'* the vertical between them is *m'l'*; the segments of the upper chord are *cn'* and *dl'*, and of the lower chord *m'r* and *k'r*. Thus the letters which represent the verticals and chord segments vary, depending upon which diagonals are under consideration.

45. Construction of Diagrams. *First Method.* By this method the diagrams are constructed for each kind of loading separately (see Figs. 1 *a*, 1 *b*, 1 *c*, 1 *d*, Plate II). The line of action of the resultant load in each case passes through the centre of curvature (*O*) of the upper chord, and is parallel to the closing side of the force polygon. The reactions for the wind diagrams are found from the condition that the resultant load and reactions must intersect at a common point. The construction is not shown.

In determining the stresses in the truss members, one set of diagonals (*e.g.* those dotted) are omitted, and the force polygons for all the joints are drawn in the usual way. The omitted set of diagonals are then dealt with, and the construction repeated. This second part of the construction only requires the drawing of the three additional lines corresponding to the three diagonals omitted in the first instance. The stress diagram thus contains the stresses in all the members of the truss, under the assumption that either diagonal in each panel is the one stressed. *E.g.* in the "wind right" diagram NL' is the stress in the vertical when the diagonals nm and $l'k'$ of the two adjacent panels are assumed to be in action, while $M'K$ is the stress in the same member when the diagonals in action are $n'm'$ and lk. The student should trace out the construction of the diagrams.

46. Determination of Maximum Stresses. The stresses are scaled off and inserted in the table (p. 39). The following combinations of loads will be considered : (1) dead load; (2) dead and snow loads; (3) dead load and wind on right side; (4) dead load and wind on left side; (5) dead and snow loads with wind right; (6) dead and snow loads with wind left. The basis for the determination of the stresses due to these various combinations of loads is the fact, that, since the diagonals are to be tension members, the one will be stressed which is in tension under the given set of loads. From the tabulated diagonal stresses we find that the diagonals in action are those given in the following table :

DIAGONALS IN ACTION.

DEAD.	D. & S.	D. & W. R.	D. & W. L.	D. S. & W. R.	D. S. & W. L.
mn	mn	mn	$m'n'$	mn	$m'n'$
neither	neither	lk	$l'k'$	lk	$l'k'$
ji	ji	$j'i'$	ji	$j'i'$	ji

We can now determine the actual stress in any member under any combination of loads. Thus to find the stress in the second vertical under the action of D.S. & W.R., we see from the table that the diagonals in action in the adjacent panels are mn and lk. The stress in the vertical is then represented by NK. Taking the algebraic sum of these stresses as given in the line nk of the table (p. 39), we find the required stress to be $-2760-1520+1000 = -3280$ lbs. The stress in this member for each combination of loads is found in a similar manner. The extreme results are recorded in columns 6 and 7 of the table. The student should determine the maximum and minimum stresses in all the members of the truss.

TABLE OF STRESSES.

MEMBER.	DEAD.	SNOW.	WIND R.	WIND L.	MAX.	MIN.	LOADS CAUSING MAX. S.
mn	− 4180	− 3450	− 4800	+ 9690	(12450) −12430	0	D.S.&W.R.
m'n'	+ 5600	+ 4600	+ 6370	−12910	(7300) − 7310	0	D.&W.L.
lk	0	0	− 7970	+ 7920	(8000) − 7970	0	W.R.
l'k'	0	0	+ 7970	− 7920	(7850) − 7920	0	W.L.
ji	− 4180	− 3450	+ 8060	− 3150	(10700) −10780	0	D.S.&W.L.
j'i'	+ 5600	+ 4600	−10730	+ 4200	(5150) − 5130	0	D.&W.R.
bo	+38980	+25720	+11050	+ 8070	(75650) +75750	+38980	D.S.&W.R.
{ cn	+31080	+21200	+12200	+ 5670	(64350) +64480	+31080	D.S.&W.R.
{ cn'	+27180	+17970	+ 7700	+14700			
{ dl	+28830	+19650	+17030	+ 870	(65500) +65510	+28830	D.S.&W.R.
{ dl'	+28830	+19650	+11310	+ 6540			
{ ej	+31080	+21200	+16960	+ 930	(69450) +69600	+31080	D.S.&W.R.
{ ej'	+27180	+17970	+24450	− 2030			
fh	+38980	+25720	+22000	− 2840	(86650) +86700	(36200) +36140	D.S.&W.R.
or	−27250	−17980	− 7690	+ 2110	(52900) −52920	(25100) −25140	D.S.&W.R.
{ mr	−25650	−16950	− 7250	+ 2000	(49900) −49850	(18750) −18800	D.S.&W.R.
{ m'r	−29370	−20000	−11490	+10570			
{ kr	−28820	−19660	−11320	+10360	(59700) −59800	(12850) −12780	D.S.&W.R.
{ k'r	−28820	−19660	−17020	+16040			
{ ir	−25650	−16950	−24430	+19130	(66700) −66700	(6650) − 6520	D.S.&W.R.
{ i'r	−29370	−20000	−17330	+16330			
hr	−27250	−17980	−25920	+20250	(71050) −71150	(7100) − 7000	D.S.&W.R.
{ om	− 5500	− 3600	− 1530	+ 400	(10500) −10630	(400) − 420	D.S.&W.R.
{ on'	− 9000	− 6550	− 5580	+ 8580			
{ nk	− 2760	− 1520	+ 1000	− 4440	(4250) − 4280	(1950) + 1950	D.&S.
{ m'l'	− 5550	− 3800	− 7710	+ 7500			
{ lj	− 2750	− 1520	− 3100	− 360	(7100) − 7080	(2300) + 2400	D.S.&W.R.
{ li'	− 5550	− 3800	+ 2270	− 2440			
{ k'j	− 2760	− 1520	− 8670	+ 5160			
{ ih	− 5500	− 3600	− 5200	+ 4070	(14000) −13950	(1350) − 1430	D.S.&W.R.
{ j'h	− 9000	− 6550	+ 1600	+ 1400			

47. Construction of Diagrams. *Second Method.* Thus far the stresses due to wind, snow, etc., have been determined separately. Another method is to draw the stress diagram for each combination of loads, in which case the maximum and minimum stresses can be taken at once from the diagrams. The latter method possesses greater advantages when dealing with trusses having counterbracing than otherwise. In the example, six diagrams would be required by the second method, as six combinations of loads are considered. The dead load diagram is not repeated. The remaining ones are given in Figs. 1 e, 1 f, 1 g, 1 h of Plate II. The same lettering is used as for the first method. Referring to Fig. 1 e, the external forces are laid off in order. At the second joint, for example, the resultant load is BC, made up of Ba, which is one-half the wind load on b; $a\beta$, one-half the wind load on c; βC, the vertical load on the joint. The line of action of the resultant load, marked D. & W., acts through the centre of curvature and parallel to AF. The reactions FR and RA are found by applying the funicular polygon to the resultant load.

In constructing the force polygons for the joints, that diagonal of each panel which is found by trial to be in tension under the given loads is used. It is of advantage to draw all the lines corresponding to the chord stresses at first. Then (Fig. 1 e), having constructed the polygon $ABOR$, we must determine which diagonal of the second panel is in action. Omitting the dotted one, we construct the polygons OMR and $BCNMO$. The order of letters indicates that mn would be in compression, and hence that $m'n'$ is the diagonal in action. The final construction is then made, using the latter diagonal. It is not necessary actually to draw the additional lines of the trial polygons. OM must be drawn, and its intersection with MR locates M. CN being already drawn, the position of M relatively to CN will indicate the kind of stress in mn. Each panel is dealt with in a similar manner. The construction of the remaining diagrams is like that explained.

Figure 1 h contains the stresses due to D.S. & W.R. and also D.S. & W.L. This combination diagram is made to save space and labor. It is evident from the symmetry of the truss that the stress in any member due to D.S. & W.L. is the same as that which would exist in the corresponding member on the other side, if the wind acted on the right side and the roller were transferred to the right end. By following this course the two wind diagrams can be combined as shown. The additional construction lines are dotted and letters underscored. For example, $\underline{MR}$ is the stress in ir, $\underline{MN}$ in ij, etc., for D.S. & W.L. loads, the roller being at the left end.

48. Maximum and Minimum Stresses. The maximum and minimum stresses can be determined at once by comparing the lengths of the corresponding lines in the various diagrams. The results, obtained independently of those previously given, are inserted in parentheses in columns 6 and 7 of the table (p. 39). The student should trace out the construction of the diagrams, and determine the maximum and minimum stresses from them.

Each of the methods described possesses advantages. The determination of maximum and minimum stresses is more laborious by the first method; on the other hand, it is desirable to know the stresses produced by the separate kinds of loads. The first method is favorable to the graphical work, as the diagrams are smaller or may be drawn to a larger scale, and are likely to be fewer in number. (See also Art. 28.)

Figure 25 N and O are other examples of trusses with counter-bracing.

§ 4. *Trusses with a Double System of Web Members.*

49. General Methods of Solution. Such a truss may be treated as a combination of two trusses having common chords but distinct systems of web members. The girder of Fig. 28 A, for example, can be resolved into those shown in Figs. 28 B and 28 C, the stresses in the members of these component trusses being found in the usual manner. The actual stress in any web member of the original truss is given directly by the diagrams, while the stress in any chord segment, as ab, is evidently equal to the algebraic sum of the stresses found for cd (Fig. 28 B) and ef

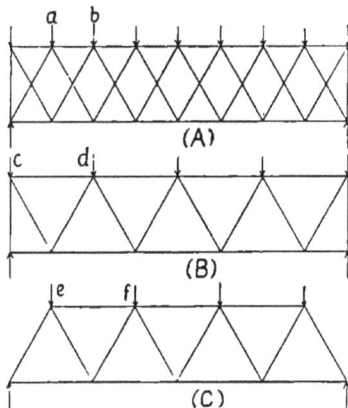

Fig. 28.

(Fig. 28 C). When the given truss can be resolved in more than one way, or when the distribution of loads between the component trusses is uncertain, the problem is indeterminate.

Figure 29 is a Crescent roof truss with two systems of web members, as shown by the full and dotted lines respectively. By tracing out the force polygons for the different joints it will be seen that the stress diagram can be drawn at once for the complete truss. A suitable system of lettering the interior of the truss is given, the intersections of the diagonals being treated as joints. If the joint

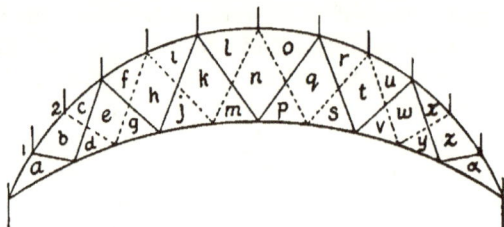

FIG. 29.

1 were made to coincide with 2, the portion of its load supported by each component truss would be uncertain and the problem in this particular would be indeterminate.

50. Double Diagonal Bracing. In the case of counterbraced panels (Fig. 1, Plate II; Fig. 25 N and O), the two diagonals are assumed not to be in action for the same loading, both being designed for the same kind of stress. When the diagonals are designed to act simultaneously (one being in tension while the other is in compression), the truss can be resolved into two trusses having common chords and verticals but distinct systems of diagonals. Each joint would belong to both component trusses, and the division of its load between the two trusses would be uncertain. One way of dealing with this case is to assume that the load at each joint is divided equally between the two trusses. The stresses in the verticals and chords would be the algebraic sum of those found for the separate trusses. The results of this method are evidently liable to err in the wrong direction.

§ 5. *Three-hinged Arch.*

51. Definition. A three-hinged arch consists of two arched ribs hinged at the crown and abutments (Fig. 30). The outward thrust at the ends may be resisted by the abutments or by a tie rod joining the ends. The ribs may be braced or solid.

52. Determination of Reactions of Hinges. The reactions are assumed to act through the centres of the hinges. The reactions of the hinge at the crown on the two half-ribs must evidently be equal and opposite. Let *ab* (Fig. 30) represent the line of action of the resultant load supported by the left half-rib, the right half being assumed to be unloaded. The reactions at O' and O'' must be equal and act along the line $O''O'$. The three forces acting on the left half must, for equilibrium, intersect at a common point. This condition determines on to be the direction of the reaction at O.

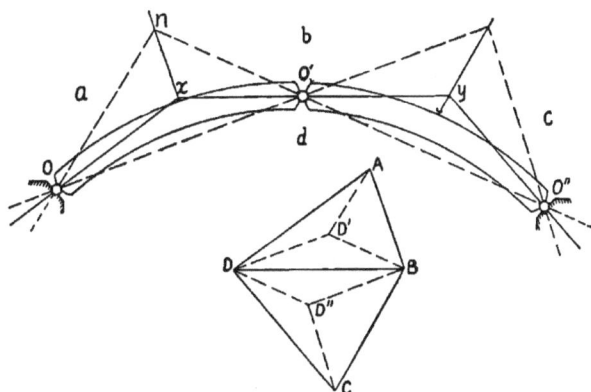

FIG. 30.

The magnitudes of these reactions can now be found by constructing the triangle of forces ABD'. BD' is the magnitude of the reaction at O'', and of the two equal and opposite reactions at O'. $D'A$ is the magnitude of the reaction at O.

The directions and magnitudes of the reactions at O, O', and O'', for any resultant load bc on the right half-rib, are found in a similar manner, the triangle of forces being BCD''. When both sides are loaded, the reaction at either hinge is evidently the resultant of the reactions due to the loads taken separately. Combining the two separate reactions for each of the hinges by the triangle of forces, we find the resultant reactions for the right half-rib to be CD and DB at O'' and O' respectively; and for the left half-rib, BD and DA at O' and O respectively.

The lines of action of these reactions are ox, $xo'y$, yo'', drawn parallel to DA, DB, and DC, respectively. These lines should intersect on ab and bc as shown, if the construction is accurate.

In the case of a symmetrical arch, symmetrically loaded, the reactions at the crown will evidently be horizontal. The reactions of the end hinges can then for this case be found directly; the points x and y being determined by the intersections of a horizontal line through O' with the resultant loads.

53. Determination of Reactions. *Second Method.* The point D (Fig. 30) can be taken to be the pole, and the lines ox, xy, yo'' the strings of a funicular polygon for the given loads. The reactions of the hinges can, then, be determined by constructing a funicular polygon for the loads, such that the three strings will pass through the hinges. The strings will be the lines of action of the reactions, and the lengths of the corresponding rays their magnitudes.

Instead of using the resultant loads, the actual loads may be employed in applying this method. In this case it is only necessary to construct a funicular polygon for the given loads, such that the three limiting strings will pass through the three hinges. (For methods, see Arts. 23 and 24.)

54. Determination of Reactions. *Algebraic Solution.* In order to secure greater accuracy, it may be desired to determine by calculation the position of the pole of the required funicular polygon. This is done by calculating the reaction at O', since the ray DB (Fig. 30) or PD (Fig. 31) represents this reaction.

Let R, R be the equal and opposite reactions at O' (Fig. 31). These are resolved into horizontal (H) and vertical (V) components.

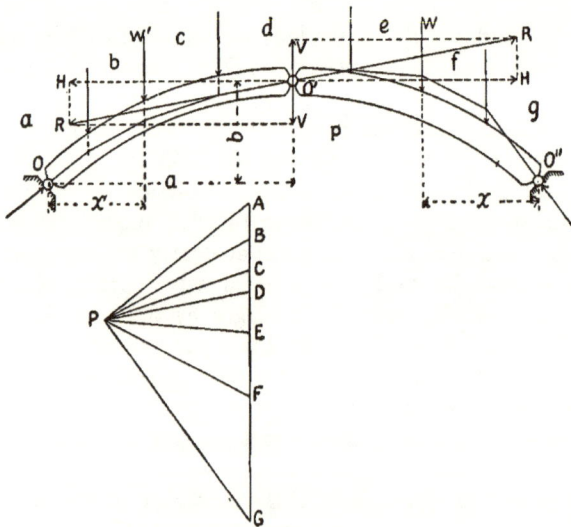

Fig. 31.

Taking the moments, about O, of the forces acting on the left half, we have $H \cdot b - V \cdot a = \Sigma W'x'$, in which W' represents any load, and x' its arm. Similarly, the moments about O'' of the forces acting on the right half give the equation $H \cdot b + V \cdot a = \Sigma Wx$. By solving these two equations, we determine the values of H and V, and consequently R. This value of R, laid off in the proper direction from the point D of the force polygon, locates the pole P of the required funicular polygon. PA and PG are then the reactions at O and O'' respectively. It is to be noted that PAD and PDG are the polygons of external forces for the two half-ribs.

PROBLEM. The semicircular arch (Fig. 32), hinged as shown, is loaded with a dead load of 8000 pounds uniformly distributed over the roof, and also with a wind load of 8000 pounds on the right side. Find the reactions of the hinges: (1) graphically; (2) by calculation, as explained. Also draw a funicular polygon for these loads, to pass through the three hinges.

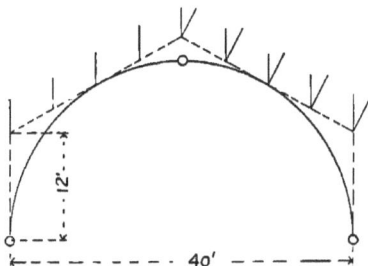

FIG. 32.

55. Determination of Stresses in Braced Arches. Method of Joints. When the two half-ribs are braced, as shown in Fig. 3, Plate III, the stresses in the members can be found by constructing the polygons of forces for the joints in succession, the methods being the same as those explained in the preceding sections of this chapter. When the number of panels is large, the treatment shown in Fig. 37, Art. 62, should be employed in order to avoid the inaccuracies resulting from a long succession of joints, and to check the work.

56. Example. (See Fig. 3, Plate III.) The general dimensions of the arch are given in the drawing. The inner curve is a semicircle of 40 ft. radius. The outer curve consists of two circular arcs of 65 ft. radius, intersecting at the apex. The dead load is taken to be 3000 lbs. per panel. In estimating the wind load, the panels were taken in pairs, the wind being assumed to act at right angles to the chord extending across two panels. The total wind pressure on the panels g' and f' is 4400 lbs.; on e' and d', 9000 lbs.; and on c' and b', 14400 lbs. From the symmetry of the arch, the maximum stresses in corresponding members on the two sides will be equal, and it is only necessary to construct the dead load diagram for one side, and the wind diagram for wind pressure on one side. The snow stress diagram is not drawn.

. *Dead Load Diagram.* The loads on the right half-arch are plotted to scale, GG' being one-half the load at the crown joint. Since the reaction of the crown hinge is horizontal (Art. 52), the pole P' is taken on a horizontal line through G, and the dotted funicular polygon (2') is constructed, the first string being drawn parallel to $P'A'$ through the right-hand hinge. The pole P of the polygon which will pass through both hinges is now located by the method of Art. 20. $G\,2$ and $G\,2'$ are laid off equal, respectively, to the ordinates of polygons (2) and (2') measured from β. Connecting 2 with P', and drawing a parallel through 2', P is located, in accordance with the condition of Art. 20. The required polygon (2) is then drawn, working from both hinges and closing half-way between the hinges. Otherwise the angles of this polygon could be located directly by the second method of Art. 24.

The reactions of the hinges being $A'P$ and PG, the polygon of external forces for the right half-arch is $PGG'F'E'D'C'B'A'P$. The stresses in the truss members can now be determined by constructing the polygons of forces for the joints in succession, beginning at the hinges. To insure greater accuracy and also check the work, the stress in the member $n'p$ of the lower chord was calculated by moments. This may be done as follows :

The string of the funicular polygon, corresponding to any panel of the truss as d', is the line of action of the resultant of all the external forces (PG, GG', $G'F'$, $F'E'$, $E'D'$) acting on either side (left) of the panel ; hence it is the line of action of the resultant external force acting on the members cut by the section $x'y'$, the magnitude of this force being the corresponding ray PD'. This force PD' is balanced by the stresses in the members $d'm'$, $m'n'$, and $n'p$. The stress in $n'p$ can then be found by taking moments about the intersection of $d'm'$ and $m'n'$, i.e. the moment of the external force PD' about this axis is balanced by the moment of the stress in $n'p$. By measurement, $PD' = 2''.56 \times 5000 = 12800$ lbs. and its arm is $1''.06$ on the drawing. The arm of the chord $n'p$ is $.94''$; hence,

$$\text{Stress in } n'p = \frac{12800 \times 1.06}{.94} = 14430 \text{ lbs. (compression).}$$

This stress is plotted in proper position in the stress diagram, and the point N' is thus located. The stresses in $d'm'$ and $m'n'$ are now determined by completing the polygon of forces $PGG'F'E'D'M'N'P$. The polygons of forces at the different joints of the arch can now be constructed in the usual way, working from the hinges and also from the panel d' in both directions.

Wind Diagram. The wind loads are plotted to scale in Fig. 3 b, and their lines of action are then drawn parallel to the proper sides of the force polygon. Assuming any pole P', a funicular polygon zz' for these loads is constructed.

The pole P of the polygon to pass through the three hinges is located as follows: Since the left side is unloaded the reaction of the crown hinge must also pass through the left-hand hinge (Art. 52). The corresponding ray PG is drawn through G parallel to this reaction, and the required pole must lie on this ray. The pole must also lie on $P\partial$, this line being located by the method of Art. 23 (the construction is shown by dotted lines). The intersection of these two lines PG and $P\partial$ locates the pole P. The funicular polygon marked (1) is constructed from this pole, the strings PG and PA' passing through the hinges as required. The polygon of external forces for the whole arch is $PGG'F'E'D'C'B'A'P$. Having determined the hinge reactions, the polygons of forces at the joints can be constructed.

To increase the accuracy of the diagram, the stresses in pn' and pn were calculated in the manner previously described for the dead load. Referring to pn, the external force (reaction of left hinge) is found to be $2''.72 \times 5000 = 13600$ lbs. Arm $3-4 = 1.87''$, arm $3-5$ $= .94''$. Taking moments about the joint (3), we have

$$\text{Stress in } np = \frac{13600 \times 1.87}{.94} = 27100 \text{ lbs.}$$

This stress is compression. By the same method we have

$$\text{Stress in } n'p = \frac{5400 \times 3.84}{.94} = 22100 \text{ lbs. (tension).}$$

These stresses were plotted in Fig. 3 b; then the force polygons for the joints were constructed as explained for Fig. 3 a. The student should trace out the construction.

57. Maximum Stresses. The maximum stress in any member, as $d'm' = dm$, is found as follows:

The dead load stress, scaled from the diagram, is

$$.34 \times 5000 = 1700 \text{ lbs. (tension).}$$

The stress in $m'd'$ for wind right is $5.53 \times 5000 = 27650$ lbs. (compression)$=$ stress in md for wind left. The stress in $m'd'$ for wind left is the same as that in md for wind right, and equals

$$2.94 \times 5000 = 14700 \text{ lbs. (tension).}$$

Comparing these results, the extreme stresses in $m'd'$ and md are found to be

$$14700 + 1700 = 16400 \text{ lbs. (tension)}$$

and $$27650 - 1700 = 25950 \text{ lbs. (compression).}$$

The funicular polygons and stress diagrams might have been constructed for the combined loadings, as explained for simple trusses. Counterbracing can also be dealt with in the same manner as previously explained.

The funicular polygon drawn through the hinges is called the *line of pressure*, since its strings represent the lines of action of the resultant pressure on the panels corresponding to these strings.

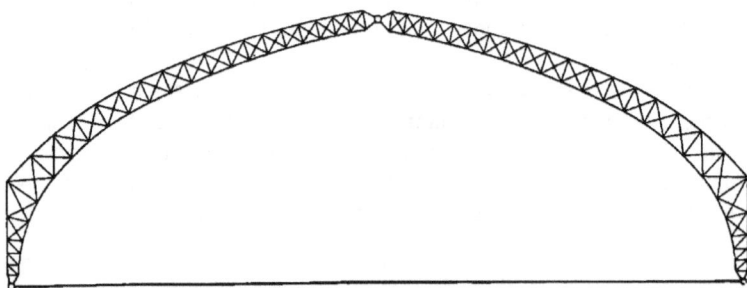

FIG. 33.

Figure 33 represents one of the arches supporting the train-shed roof of the Philadelphia and Reading Terminal Railway. The diagonals are tension members.[1]

58. Three-hinged Arch. Solid Ribs. Determination of Stresses.

Let Fig. 34 represent such an arch, the funicular and force polygons for the given system of loads being drawn. The resultant external force acting at a cross section n is $PA = F$, its line of action being the corresponding string. This force can be resolved into F'' and F''', acting respectively at right angles and parallel to the plane of the section. The latter is the shearing force at the section. The former (compounded with two equal and opposite forces F'' acting through n, the centre of gravity of the section) is equivalent to a force F'' acting at the centre of gravity and causing a uniformly distributed compression stress, and a couple whose arm is nn'. This couple causes, as in a beam, a bending stress at the section, the bending moment being $F'' \times nn'$, or its equal $F \times ny$. The bending stress

---

[1] For description, see *Trans. Am. Soc. C. E.*, 1895.

is combined with the direct compression stress, as in case of a strut loaded eccentrically. The stress at any other section is found in a similar way.

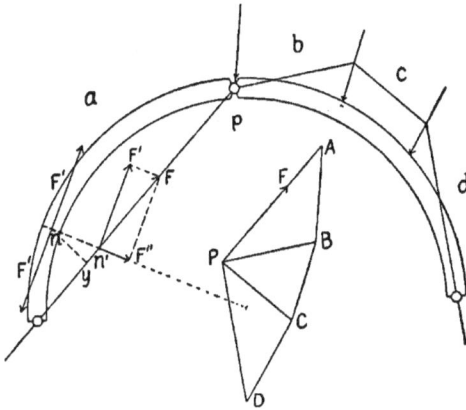

FIG. 34.

The determination of the stresses in arches fixed at the ends, and in arches hinged only at the ends, requires the use of the theory of elasticity. For the treatment of such cases, see the list of references, p. 77.

§ 6. *Trusses loaded at Other Points than the Joints. Bending Stresses.*

59. Determination of Joint Loads. The purlins are frequently supported at other points than the joints. The design of the roof may require the purlins to be placed closer together than the joints of the upper chord, *e.g.* when the roof covering, as slate or corrugated iron, is supported directly by the purlins. In such cases the upper chord is subjected to a bending as well as a direct stress, and this must be provided for in proportioning this member. The lower chord also may be subjected to bending stresses from shafting or other loads supported by it at other points than the joints, as well as from its own weight.

EXAMPLE. Let the total weight of roof covering and purlins (Fig. 35 A) be 3000 lbs. The weights at the points of support of the purlins will then be those given. Consider the load at *n*. (See Fig. 35 B.) This load can be resolved into 167 lbs. at *A* and 333 lbs. at *B*. If, now, we apply at *A* and *B* equal and opposite forces equal to the components, these will balance and not affect the stresses in the truss. The upward forces at *A* and *B* correspond to the

reactions of a beam under the load of 500 lbs., and there remain the
loads, 167 lbs. at A and 333 lbs. at B. This 500 lb. load is, therefore, equivalent to the component loads at the adjacent joints, and
in addition causes a bending stress in AB, the maximum bending
moment being at the load and equal to 167 × 5 ft. lbs. The joint

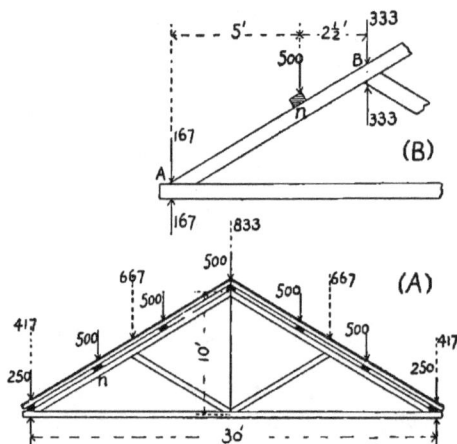

FIG. 35.

loads found as indicated are given in Fig. 35 A. All other loads
supported by the purlins are treated similarly. This method is
evidently approximate when the chord is continuous, as is commonly
the case, and when the loading is oblique to the length of the chord.

60. Combined Stresses. In Fig. 36 let the piece be subjected
to a transverse load W and a direct force P, acting through the
centre of gravity of the end sec-
tions. Let the resulting deflection
be v. We have then: (1) a bending stress due to W; (2) a bending
stress due to the eccentricity v of
P; (3) a direct compression stress.

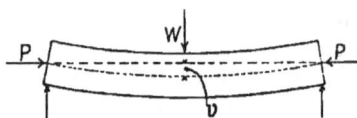

FIG. 36.

When the deflection v is sufficiently small, the bending stress (2)
can be neglected, as is commonly done. In this case the resultant
stress is the algebraic sum of the direct stress and the bending
stress due to W, i.e. $f = \dfrac{P}{A} \pm \dfrac{M}{I} y$ in which M is the bending moment
due to W.

If the force P does not act through the centre of gravity of the end sections, there results an additional bending stress due to this eccentricity.

§ 7. General Directions.

61. Instruments. In order to secure a satisfactory degree of accuracy in graphical solutions, special precautions must be taken. In general, the drawings should be made with such care and in such a manner as to leave no question as to their accuracy at any stage of the construction.

The wooden edges of the ordinary drawing board and T-square are unreliable. A steel straight edge with lead weights to hold it in position is preferable. The edges of the triangles must be straight and the 90° angle true. The usual hard rubber and celluloid triangles are not sufficiently accurate in these respects and should be tested before using. The best quality of dividers and compasses must be used. A metal scale graduated to hundredths of an inch and a very hard pencil kept sharpened to a fine chisel edge, complete the list of necessary instruments. All intersections to be preserved are located accurately by a fine prick point enclosed in a circle.

62. Construction of Diagrams. The space diagram is constructed accurately and, if the directions of the lines of the stress diagram are to be taken from it, to a large scale. In constructing the stress diagram the directions of its lines are commonly taken from the space diagram. In doing this, a good general rule to follow is not to obtain the direction of a line of the stress diagram from a shorter line of the truss. In order to secure accuracy in this respect without constructing the truss to an extravagantly large scale, the construction may sometimes be made by indirect means.

ILLUSTRATION. The direction of BS (Fig. 1 a, Plate I) may be obtained by laying it off to the same slope as the rafter (*i.e.* making $RS = \frac{40}{16} BR$), instead of drawing it parallel to that member in Fig. 1. CU, DW, etc., are drawn parallel to BS. ST, UV, etc., are drawn perpendicular to the base line. TU, VW, etc., are drawn to the same slope as the members of the truss. Thus the whole stress diagram might be constructed without taking the direction of any line from the truss diagram. When the chords are arcs of circles, the chord stresses should be drawn at right angles to the radius bisecting the chord member.

The intersections of the lines of the stress diagram should be

located accurately.   If the slope of the rafter is small, the angle
between it and the lower chord is very acute and the intersection
of the corresponding lines *BS* and *RS* (Fig. 1, Plate I) is liable to
considerable error.   As the accuracy of the remainder of the work
depends upon the correct location of the point *S*, it may be desirable
to locate it by calculation in some such way as explained in Art. 36.

In important work it is well to determine the supporting forces
by calculation, as the correctness of all the stresses depends on
them.   All such calculations serve as checks on the graphical
work.

As the construction proceeds from joint to joint of the truss,
errors accumulate ; hence a long succession of joints in constructing the stress diagram should be avoided.   In a truss of comparatively few panels it is sufficient to work from each end towards the
centre.   In case of a large number of panels, the truss should be
broken up into sections, using the method of sections (Art. 6, Third
Solution).   In Fig. 37, for example, we can find the stress in *mn* by
moments and lay off the result *MN* in the stress diagram.   By the

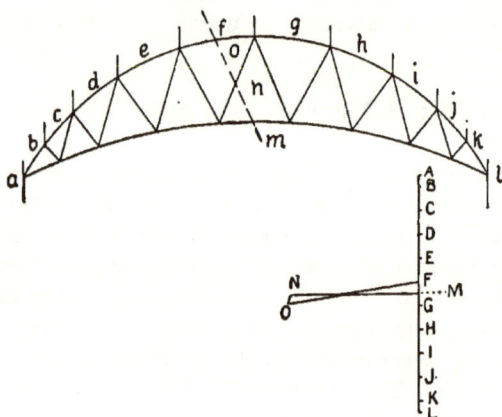

Fɪɢ. 37

polygon of forces we then determine *FO* and *ON* to be the stresses
in these members.   We can now complete the diagram, working
from each end towards the centre and from the centre towards
the ends, closing half-way between centre and ends.   In the same
manner the truss may be divided into any number of sections.

63. Checks.   The accuracy of the construction is checked by the
diagrams closing.   This, however, does not insure the correctness

of the results, as the truss diagram may have been drawn incorrectly or errors made in plotting the loads. It is well, in addition, to check any important work by computing the stresses in one or more of the members by moments or otherwise. When the computations by moments are made at the outset, and the results introduced into the diagrams, as indicated in the preceding paragraph, the proper closing of the diagrams is equivalent to both the checks mentioned.

64. Scale of Stress Diagram. Nothing is gained by making this scale excessively large. In small trusses it can always be made large enough without making unwieldy diagrams. The lines of the stress diagrams, when drawn with the precautions described above, . need not be in error more than two or three hundredths of an inch. With a scale of 10000 lbs. to the inch, this would be an error of 200 to 300 lbs. in the stress. Such an error is of no importance in large work requiring so small a scale.

The student is recommended to study trusses with the view of understanding the purpose and action of their various members independently of the stress diagrams. In Fig. 24, for example, the load cd tends to deflect the rafter, this deflection being prevented by the brace kj, which would therefore be in compression. kj, being in compression, exerts a downward thrust on the lower chord at a. The lower chord is prevented from deflecting under this thrust by ji; ji is therefore in tension. A study of trusses in such a manner will assist the student to a better understanding of their design.

CHAPTER III.

BEAMS.

§ 1. *Shearing Force and Bending Moment.*

65. Definitions. The *shearing force* at any section of a beam is equal to the algebraic sum of the external forces to the left of the section.

The *bending moment* at any section of a beam is equal to the algebraic sum of the moments of the external forces to the left of the section, the moment axis being the neutral axis of the section.

It was shown in Art. 18 that if an ordinate be drawn at any section of a beam, the distance intercepted on this ordinate by the funicular polygon, multiplied by the pole distance, is equal to the bending moment at the section. The scale of the ordinate is the same as that of the space diagram, and the scale of the pole distance is the same as that of the force polygon.

66. Beam supported at Ends and loaded with Concentrated Loads. The force and funicular polygons (Fig. 38) are drawn, and the sup-

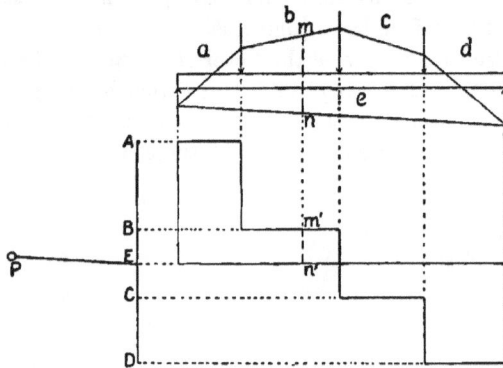

Fig. 38.

porting forces *DE* and *EA* determined by means of them. (See Art. 12.) By Art. 18, the bending moment at any section, as *mn*,

54

is equal to the intercept *mn*, times the pole distance. The intercept *m'n'*, in the lower diagram, represents the shearing force at the section. This shear diagram is constructed directly from the definition of shearing force, and needs no description. It will be seen by inspection of these two diagrams that the maximum shearing force is at one end and the maximum bending moment at one of the loads, *bc*.

67. Overhanging Beam. (Fig. 39.) The construction will be understood without explanation. The shear diagram is omitted. It

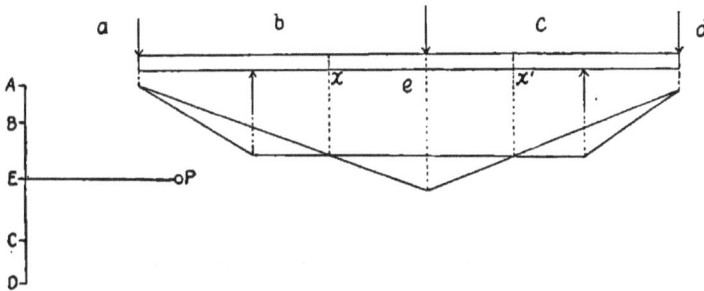

Fig. 39.

will be noticed that the bending moments at *x* and *x'* are each zero. These points are therefore the points of inflexion of the elastic curve which the beam will assume under the given loads. The bending moment between *x* and *x'* is positive, and the elastic curve is convex downward. Outside of these points the moment is negative, and the curve is consequently convex upward.

68. Distributed Loads. (Fig. 40.) Let the load be divided as shown by the broken lines, and each division be treated as concentrated at its centre of gravity. These concentrated loads can be dealt with as indicated in the preceding cases. The funicular polygon for them is drawn. The bending moments at the points of division of the load are the same as those caused by the concentrated loads, and are therefore correctly given by the intercepts of the funicular polygon. The exact diagram for the loaded portion of the beam is then the inscribed curve, the points of tangency, corresponding to the points of division of the load, being indicated by circles. For a uniformly distributed load, this curve is a parabola. The shear diagram needs no explanation.

If the bending moment at any given section of the beam is desired, it can be found by making this section a point of division of the load. Thus the ordinate nn' (Fig. 41) represents the bending moment at that section for a load uniformly distributed over the

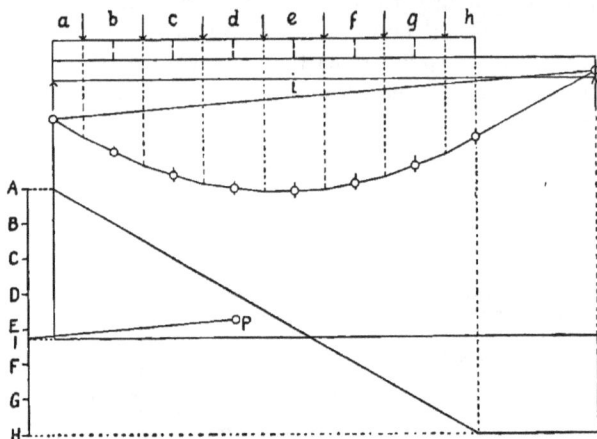

Fig. 40.

entire beam, the funicular polygon being constructed by dealing with the loads on the two sides of this section as concentrated at

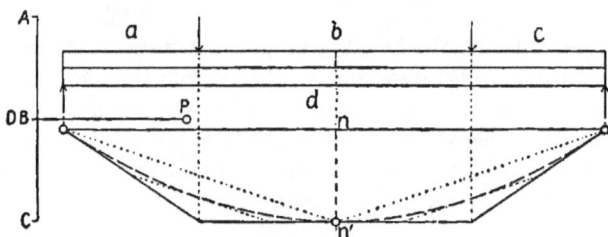

Fig. 41.

their centres of gravity. The true diagram is the parabola drawn in broken lines tangent to the strings b, a, and c, at n' and the ends of the beam.

69. Problems. 1. Assume a beam loaded with given concentrated loads. Find the bending moment at any section and check the result by calculation. Also determine the maximum bending moment.

2. Assume a beam loaded with a given uniformly distributed load extending over part of the span. Find the bending moment at any section and check the result by calculation.

3. In Prob. 2 draw the moment curve and determine the value of the maximum bending moment. Check the result by calculation.

§ 2. *Deflection of Beams.*

70. Graphical Determination of Elastic Curve. In simple cases the usual formulas furnish the best method for determining the deflection of a beam. For beams of non-uniform section, or unusual system of loads, the deduction of the formulas becomes tedious. In such cases a graphical solution can be used to advantage when extreme accuracy is not required.

EXAMPLE. The beam (Fig. 42), of uniform section, is supported at the ends and loaded with two concentrated loads as shown. It is required to construct the elastic curve.

Solution. Construct the funicular polygon (Fig. 42 A) for the given loads. The force diagram is Fig. 42 A'. Treat the surface of

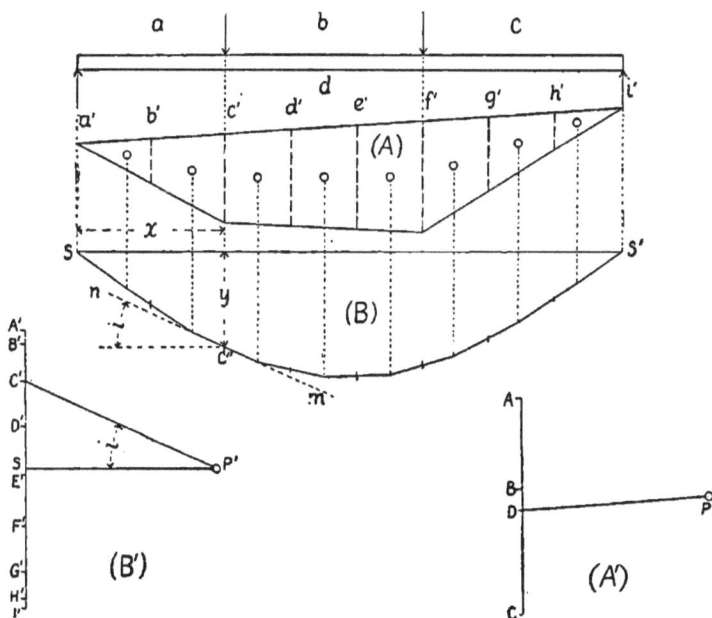

FIG. 42.

this polygon as if it represented a distributed load for the same beam, and construct a second funicular polygon. For this purpose the surface (Fig. 42 A) is subdivided by ordinates, and each division is concentrated at its centre of gravity, marked by a circle. The areas of these surfaces are plotted to scale in the force diagram (Fig. 42 B'), and the corresponding funicular polygon (Fig. 42 B) is drawn. The pole P' is taken so that SS' is horizontal. The exact diagram will be a curve inscribed in this polygon (Art. 68). This curve is the elastic curve of the beam, the deflection at any point being represented to a known scale by the length of the intercept between the curve and the line SS', this line being drawn so as to satisfy the condition that the deflection at each support is zero.

Proof.   Let $O =$ any ordinate of Fig. 42 A (in., full size).

Let $D =$ pole distance of Fig. 42 A$'$ (lbs.).

Let $D' =$ pole distance of Fig. 42 B$'$ (sq. in., full size).

Let $M =$ bending moment at any section of the beam (in. lbs.).

Take the origin at the left support, and let C' be any section of the beam whose distance from the origin is x.

The ordinate y (Fig. 42 B) at the given section is intercepted by SS' and mn, the latter being the tangent to the funicular curve at C''. Draw the rays $P'S$ and $P'C'$ parallel to these lines. The angle $C'P'S$ is then equal to the angle of slope i of the tangent mn.

Hence,
$$\tan i = \frac{dy}{dx} = \frac{SC'}{P'S} = \frac{SA' - A'C'}{P'S = D'} \quad \cdots \quad \cdots \quad (1)$$

In Eq. 1 SA' is constant and $A'C'$ is equal to the area in Fig. 42 A, lying to the left of the section being considered; *i.e.*

$$A'C' = \int_0^x O dx.$$

Substituting in Eq. 1,

$$\frac{dy}{dx} = \frac{SA' - \int_0^x O dx}{D'}.$$

Differentiating and dividing by dx, we have, neglecting signs,

$$\frac{d^2 y}{dx^2} = \frac{O}{D'}. \quad \cdots \quad \cdots \quad \cdots \quad \cdots \quad (2)$$

The general equation of the elastic curve (Lanza, p. 301) is

$$\frac{d^2 y}{dx^2} = \frac{M}{EI} \quad \cdots \quad \cdots \quad \cdots \quad \cdots \quad (3)$$

In order for the curve (Fig. 42 B) to be the elastic curve of the beam, the second members of Eqs. 2 and 3 must be equal.

$$\therefore \frac{O}{D'} = \frac{M}{EI} = \frac{O \cdot D}{EI}. \quad \therefore DD' = EI \cdot \quad \cdots \quad \cdots \quad (4)$$

Let a = ratio of true to plotted dimensions of the beam. Then, in order for the ordinates of Fig. 42 B to represent the true deflection of the beam, the vertical scale of Fig. 42 B must be increased a times. For this purpose D' must be reduced in the same ratio (Art. 20); also, if we wish to exaggerate the deflections in any ratio n, D' must be reduced proportionally. In this latter case Eq. 4 becomes

$$DD' = \frac{EI}{na} \quad \cdots \cdots \cdots \quad (5)$$

By means of Eq. 5 suitable values of D and D' can be determined such that the ordinates of the second funicular polygon (Fig. 42 B) shall represent the deflection of the beam exaggerated n times.

It is not necessary for SS' (Fig. 42 B) to be horizontal, since the ordinates are constant so long as the pole distance D' is not altered.

If the bending moment changes sign as in Fig. 39, the areas corresponding to negative moments must be plotted in Fig. 42 B' in the opposite direction from the positive ones.

If the beam is of *non-uniform section*, Eq. 5 shows that DD' must vary in the same ratio as I. The proper elastic curve for this case is then constructed by making the pole distance D' vary in the same ratio as I varies.

To determine the deflection at any given section of the beam, make the section one division line of the area (Fig. 42 A). In other respects the area may be subdivided in any manner (see Art. 68), with the exception that when the beam is of non-uniform section the area must be divided at those sections where the value of I changes.

71. Example 1. Fig. 1, Plate III, is a cantilever, 10 ft. span, loaded at two points. I has three different values, as given in the diagram. $E = 30000000$ lbs. per sq. in. It is required to draw the moment diagram and the elastic curve.

Scale of space diagram, $1 : 20$. $\therefore a = 20$. Scale of force diagram (Fig. 1 a'), 4000 lbs. $= 1''$. The pole distance PA is taken to be $4''$. $\therefore D = 4 \times 4000 = 16000$. The construction of the moment diagram (Fig. 1 a) needs no explanation. Its surface is divided as shown and the centres of gravity are indicated by circles. The areas of these divisions are plotted in Fig. 1 b' to the scale, 1000 sq. in. (full size) $= 1''$. The full size area is obtained by multiplying the diagram area by $a^2 = (20)^2 = 400$. The deflections are to be exaggerated 5 times, *i.e.* $n = 5$.

Substituting the preceding values in the formula $DD' = \dfrac{EI}{na}$, we have $16000\,D' = \dfrac{30000000 \cdot 200}{5 \cdot 20}$. $\therefore D' = 3750 = 3.75''$ to the scale (1000 sq. in. $= 1''$). The pole P' is taken with a pole distance of $3.75''$, and the elastic curve (Fig. 1 b) is constructed. The strings corresponding to $P'G'$, $P'F'$, $P'E'$ are drawn in order. At the section e' the value of I changes to 150, and the pole distance must be reduced in the same ratio, the new pole P'' lying on $P'E'$. The construction is continued in a similar manner, the pole for the portion of the beam where $I = 100$ being P'''. The ordinate mn, divided by 5, determines the deflection of the beam to be $.496''$. The computed deflection is $.493''$.

EXAMPLE 2. The beam (Fig. 2, Plate III) is supported at two points, a and h, 14 ft. apart, and overhangs 6 ft. at the right end. It is loaded with a distributed load of 300 lbs. per foot. Cross section $= 6'' \times 12''$. $I = 864$. $E = 1200000$. Scale, 1 : 40.

The load is divided into 10 equal lengths, each division being concentrated at its middle point. Scale of force diagram (Fig. 2 a'), 2000 lbs. $= 1''$. Pole distance $= 1\frac{1}{2}''$; $i.e.$ $D = 3000$. The surface of the moment diagram (Fig. 2 a) is divided as shown and the areas plotted in Fig. 2 b' to the scale, 500 sq. in. $= 1''$, the positive and negative areas being laid off in opposite directions, $n = 5$.

Substituting in the formula $DD' = \dfrac{EI}{na}$, we have

$$3000\,D' = \frac{1200000 \cdot 864}{5 \cdot 40} = 1728 \,;$$

$i.e.$ the pole distance (Fig. 2 b') is $\frac{1728}{500} = 3''.456$. Fig. 2 b is the elastic curve. The measured ordinates at 7 ft. and 20 ft. from the left end divided by 5, give for the deflections at these points $- .144''$ and $+ .012''$. The computed deflections are $- .14''$ and $+ .01''$.

§ 3. *Beams supported at More than Two Points.*

72. Determination of Reactions. In determining the reactions of the supports for such cases, the theory of elasticity must be employed, no solution being possible by means of the methods of statics of rigid bodies. From the theory of beams we know that the deflection at any point due to any system of loads is equal to the algebraic sum of the deflections produced by the loads taken separately. A method of determining the supporting forces, based on this fact, is illustrated by the following example:

The beam (Fig. 43) of uniform section is supported at four equidistant points, all on the same level, as shown, and loaded uniformly, the total load being W. It is required to find the reactions of the supports.

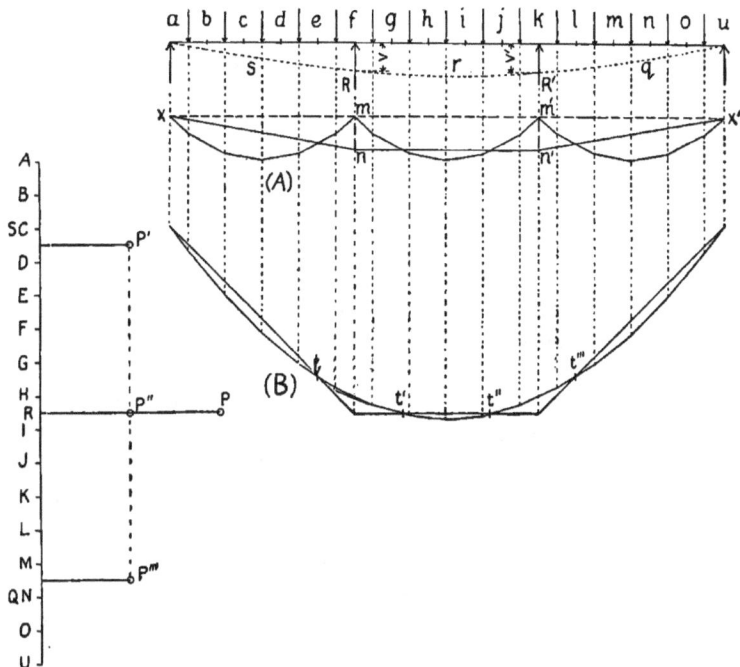

FIG. 43.

If the intermediate supports R and R' were removed, the beam would deflect as shown by the dotted curve. The deflections at the supports would then be $v = v' = \dfrac{11\,Wl^3}{972\,EI}$ (Lanza, p. 303, Eq. 2). The reactions R and R' must evidently be equal to the forces which would raise these points to their original positions, i.e. would produce upward deflections at R and R' equal to v and v' respectively.

The deflections at the intermediate supports due to the force R are $v_R = \dfrac{4\,Rl^3}{243\,EI}$ and $v'_R = \dfrac{7\,Rl^3}{486\,EI}$ (Lanza, p. 308, Eq. 2). The deflections at these points due to R' are

$$v_{R'} = \frac{7\,R'l^3}{486\,EI} \quad \text{and} \quad v'_{R'} = \frac{4\,R'l^3}{243\,EI}.$$

Hence, $v = v_R + v_{R'} = \dfrac{4\,Rl^3}{243\,EI} + \dfrac{7\,R'l^3}{486\,EI} = \dfrac{11\,Wl^3}{972\,EI}.$

Similarly, $v' = v'_R + v'_{R'} = \dfrac{7\,Rl^3}{486\,EI} + \dfrac{4\,R'l^3}{243\,EI} = \dfrac{11\,Wl^3}{972\,EI}.$

Solving for R and R', we obtain $R = R' = \frac{11}{30} W$. The two remaining supporting forces can now be found by moments. In this case each equals $\frac{4}{30} W$. The general method of determining supporting forces, illustrated by the preceding example, is of wide application, but in any complicated case is laborious.

73. Construction of Moment Diagram. The load W is divided into 15 equal divisions. These loads and the reactions previously found are plotted in order in the force diagram, and with P as pole, the moment diagram (Fig. 43 B) is constructed. The moments are negative in the spaces t–t' and t''–t''', these four points being the points of inflexion of the elastic curve.

This diagram can better be constructed as follows: So long as the pole distance remains constant, the intercepts of Fig. 43 B will not be altered; hence different poles can be used in constructing different parts of the funicular polygon, provided the pole distances are all equal. In Fig. 43 A the moment diagram is constructed, using the poles, P', P'', P''', for the strings lying in the three spans respectively. The pole distances are reduced one-half, hence the intercepts will be double those of Fig. 43 B (Art. 20). This use of different poles gives a much more satisfactory diagram than when a single pole is used.

74. Second Method of constructing Moment Diagram and Determining Reactions. Referring to Fig. 43 A, the funicular polygon $xmm'x'$ for the loads can be drawn without knowing the supporting forces. In order to complete the moment diagram it is necessary to know the values of the intercepts mn and $m'n'$. These intercepts are equal to the bending moments at these points divided by the pole distance (Art. 18). Hence the points n, n' can be located and the diagram completed when the bending moments at the intermediate supports are known. These moments can be computed by means of the "three-moment equation." (For the method of determining the bending moments at the supports of a continuous girder by means of the "three-moment equation," see Lanza's *Applied Mechanics*, Chap. VIII.) Having completed the moment diagram by drawing the strings xn, nn', and $n'x'$, the reactions UQ, QR, RS, SA, are determined by drawing the rays $P'S$, $P''R$, $P'''Q$, parallel to these strings.

From the moment diagram, the bending moment at any point can be determined, and the elastic curve constructed in the manner previously described.

The values of the reactions of the supports (or bending moments at the supports) for the case of a continuous beam of uniform section, equal spans, and loaded uniformly over its entire length, will be found in text-books on the subject. The stresses in continuous beams are quite uncertain, as they are altered by unequal settling of the supports.

75. Centre of Gravity. The following constructions for centre of gravity will be found useful in connection with the work of this and the following chapter.

1. *Centre of Gravity of Any Quadrilateral.* Two constructions are given in Fig. 44: (1) Bisect each diagonal. Connect the point of intersection of the diagonals with the middle of the line DE, joining their middle points. The centre of gravity will be at the extremity of this line when extended one-third of its

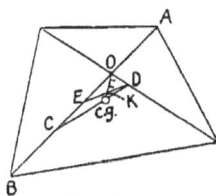

FIG. 44.

length. (2) Lay off $BC = OA$ and join C with the middle, D, of the other diagonal. The centre of gravity lies on this line one-third of CD from D.

2. *Centre of Gravity of a Trapezoid.* Two constructions are given in Fig. 45: (1) Draw the medial line GH and one diagonal. Bisect the medial line at K. The distance from the point of intersection L to K, prolonged one-third of its length, will locate the centre of gravity. (2) Lay off $AF = CD$, and $DE = AB$. The intersection of EF and the medial line is the centre of gravity.

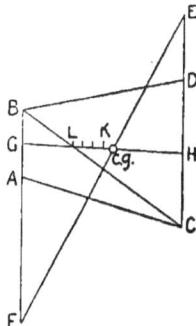

FIG. 45.

CHAPTER IV.

MASONRY ARCHES, ABUTMENTS, ETC.

§ 1. *General Conditions of Stability.*

76. Nature of the Forces involved. Let *PRMN* (Fig. 46) be a block of masonry acted upon by a force *AB*. In addition to this force, the weight of the block must be taken into account. This weight is represented by *BC*, its line of action *bc* being drawn through the centre of gravity of the block. The resultant of these two forces is *AC*, its line of action passing through the intersection of *ab* and *bc*. *AC* is, therefore, the resultant pressure exerted by the block upon the plane *MN*. This plane may be taken to be a

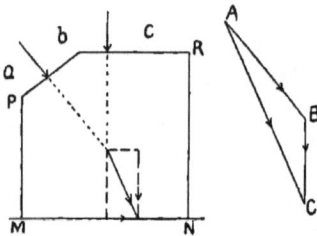

FIG. 46.

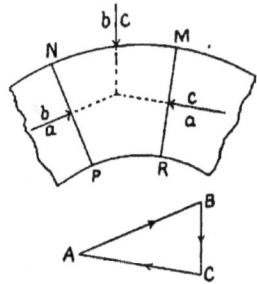

FIG. 47.

joint of the masonry or its base. Also, the forces which hold the block in equilibrium are *AB*, *BC*, and the reaction of the plane *MN*, this last being a force equal and opposite to *AC*. Moreover, *AC* represents the resultant stress on the plane *MN*.

Again, let *PRMN* (Fig. 47) be an arch stone, *BC* being the load supported by this stone including its own weight. The line of action *bc* of this load passes through the centre of gravity. *BC* is balanced by the forces *AB* and *CA* exerted upon *PRMN* by the adjacent arch stones. These three forces must, therefore, form a triangle, and their lines of action must intersect at the same point.

77. Resistance of a Masonry Joint. The conditions of stability for the block of Fig. 46 as far as the joint MN is concerned are evidently the following: (1) the block must not overturn about an edge, as N; (2) it must not slide over the joint; (3) the material of the stone and mortar must not crush. These three conditions will be discussed in turn.

78. Resistance to Overturning. In this connection, the tensile strength of the mortar joint is commonly neglected. Then the block (Fig. 46) would evidently overturn, if the line of action of the resultant force AC pierced the plane MN outside of the surface of the joint. The moment of AC about N as moment axis is the measure of the resistance to overturning about this edge, *i.e.* in order to overturn the block it would be necessary to apply a force whose moment about N was equal to that of AC, but having the opposite sign.

79. Resistance to Sliding. Let the resultant force AC (Fig. 46) be resolved into components parallel and perpendicular to the joint, as indicated. The normal component represents the direct pressure on the joint, while the parallel component tends to slide the block over the joint, and must be resisted by the sliding friction at the joint, the adhesion between the stone and mortar being neglected.

Coefficient of Friction. Let P (Fig. 48) be the resultant pressure of the block on the plane AB, and ϕ the minimum angle of inclination with the normal at which sliding will occur. This angle ϕ is called the angle of repose, and tan ϕ, or ratio of tangential to normal component of the force, is called the coefficient of friction f. It is shown by experiment that f is practically constant for given surfaces, *i.e.* is independent of the intensity of the normal pressure. In the case of masonry joints, the minimum value of the coefficient of friction is taken to be from .4 to .5. In order, then, for sliding not to occur, the resultant pressure at any joint must make with the normal an angle less than $\tan^{-1}$ .4.

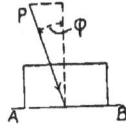

Fig. 48.

80. Resistance to Crushing. The normal component of AC (Fig. 46) represents the resultant compression stress at the joint MN. This stress is assumed to be uniformly varying. (See Lanza, p. 265.) The three cases which may occur are represented in Fig. 49. In Fig. 49 A the stress is distributed over the whole surface of the joint, the limiting case being Fig. 49 B, where the neutral axis is at one edge M. In Fig. 49 C the pressure is distributed over the portion XN of the joint. If the joint were capable of resisting tension, this last would be the case where the stress is partly tension and partly compression, but, assuming the joint incapable of resisting tension, the portion MX is without stress and tends to open.

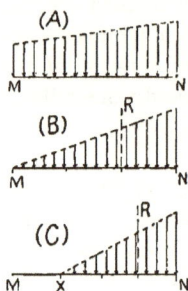

(A)

(B)

(C)

FIG. 49.

If the surface of the joint is rectangular, the resultant stress R, in Fig. 49 B, acts at $\frac{2}{3} MN$ from M. *Proof:*

Let $MN = h = $ depth of joint.

In the formula $x_1 = \dfrac{I}{x_0 A}$ (Lanza, p. 265) we have

$$I = \tfrac{1}{3} bh^3 = \tfrac{1}{3} Ah^2, \quad x_0 = \tfrac{1}{2} h.$$

Substituting, $x_1 = \dfrac{\tfrac{1}{3} Ah^2}{\tfrac{1}{2} Ah} = \tfrac{2}{3} h.$

In Fig. 49 A the resultant stress lies nearer the middle of the joint than in Fig. 49 B; hence, *In order for the pressure to be distributed over the whole surface of a rectangular joint, the resultant must act within the middle third of its depth.* The corresponding limits within which the resultant pressure must act, in case of any other form of surface, can be found in a similar manner.

When the stress is distributed over the whole surface of the joint, its maximum intensity can be found by the formula for short struts (Lanza, § 207). When the stress extends over a portion of the joint, as in Fig. 49 C, the short strut formula is evidently inapplicable. In such a case, when the surface is rectangular, the point X can be found by making $NX = 3 NR$, R being the point of application of the resultant stress. Its maximum intensity is then equal to $\dfrac{2 R}{\text{area } NX}$. If R acts at the edge N, the area NX becomes zero and the intensity of the stress infinity.

For safety the maximum stress must not exceed the working compression strength of the material.

§ 2. *Masonry Arch. Line of Pressure.*

81. Definitions. In Fig. 50, a and b are respectively the *span* and *rise* of the arch, h is the *thickness* of the arch ring, also the *depth* of the joints of the arch ring. The highest part of the arch is the

crown. The portions of the arch ring between the crown and abutments are the *haunches.* The arch stones are also called *voussoirs.*
The inner surface, *cmc*, of the arch ring is the *intrados.* The outer
surface, *dm'd*, is the *extrados.* These names are also given to the

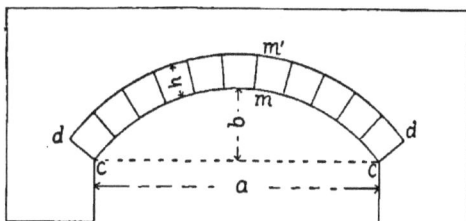

Fig. 50.

corresponding curves. The lines *c, c,* where the intrados meets the
abutments, are the *springing lines.*

Arches are designated according to the form of the intrados. A
full centred arch is one whose intrados is a semicircle. A *segmental*
arch is one whose intrados is a circular arc of less than 180°. A
pointed arch is one where the two half-curves of the intrados intersect at the crown instead of forming a continuous curve. The
wooden frame which supports the arch during construction is the
centre. (For further definitions see Baker's "Masonry Construction.")

82. Line of Pressure, a Funicular Polygon. Let AB, BC, CD,
etc. (Fig. 51), be the loads supported by the arch stones. If the
pressure at any joint, as a, is given
completely, the pressures at the other
joints can be found by the triangle of
forces (Art. 76). Thus, representing
the pressure on the joint a by PA, the
resultant of PA and AB, *i.e.* PB, will
be the pressure on the joint b, its line
of action passing through the intersection of pa and ab, as shown. Similarly, PC is the pressure on the joint c,
PD on the joint d, etc. Thus it is
seen that the lines of action of the
resultant pressures on the succes-

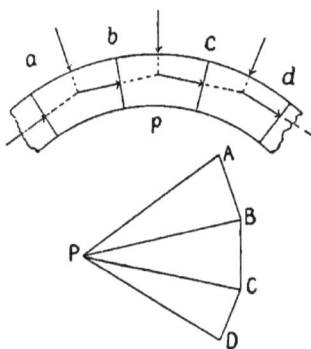

Fig. 51.

sive joints of an arch are the strings of a funicular polygon, the
corresponding rays representing the magnitudes of these pressures.

This funicular polygon will be referred to as the *line of pressure* of the arch, although the *line of pressure* or *line of resistance*, as commonly defined, *is the broken line joining the centres of pressure of the successive joints.*

83. Symmetrical Arch. Symmetrically loaded. In this case it is evident that the line of pressure will also be symmetrical with reference to a vertical through the crown, and hence the pressure at the crown will be horizontal. Only one-half the arch ring need then be considered.

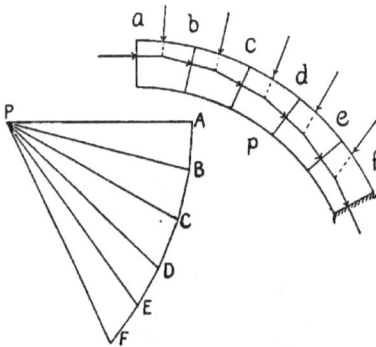

Let *PA* (Fig. 52) be the horizontal crown pressure, *AB*, *BC*, etc., being the loads. The funicular polygon constructed by using *P* as pole is then the line of pressure for the half-arch.

FIG. 52.

84. Test of Stability. A common test for the stability of a proposed arch, as far as its outline is concerned, is to determine the possibility of drawing a line of pressure which will lie wholly within the middle third of the thickness of the arch ring. If the resultant pressure at each joint acts within the middle third of its depth, the compression stress will be distributed over the whole surface of the joint (Art. 80). The general method of determining the possibility of drawing a line of pressure within the middle third of the arch ring is illustrated in Fig. 53.

The arch is full centred. The two curves drawn include between them the middle third of the thickness of the arch ring. The loads *AB*, *BC*, etc., are taken to be vertical. Assume any pole *P* on a horizontal line through *A* and construct the funicular polygon *mn*. Next assume two points, as *X* and *Y*, in the arch ring, and construct a funicular polygon to pass through them. To do this we know that the reactions at any two joints, as *X* and *Y*, must balance the resultant load included between these joints. This resultant load *R* acts through the intersection of the strings *a* and *g* of the polygon *mn*. The corresponding strings of the desired polygon must then intersect on *R*. The string *a*, passing through *X*, is horizontal; the string *g*, therefore, has the direction *YO*. The pole *P'* of this polygon is now located by drawing the ray GP' parallel to YO, and the polygon is constructed. This polygon XY departs farthest from the middle third of the arch ring at the joint e. A polygon whose strings a and e pass through X and Y' respectively is then most

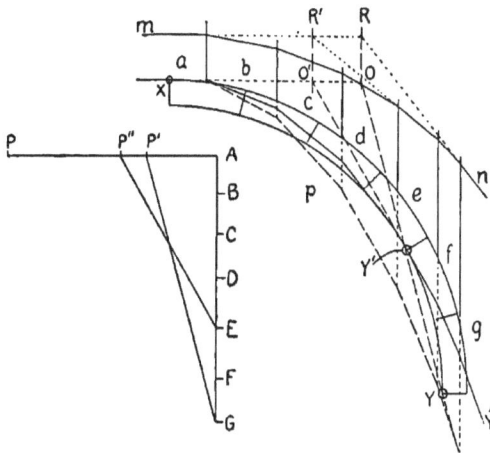

Fig. 53.

likely to fall within the required limits. The pole P'' of this polygon is located in a manner similar to P'. This last polygon $XY'Y''$ lies within the designated limits except near the springing plane, where the requirement is not essential in the case of full centred arches, since this portion of such an arch can be treated as part of the abutment.

85. Maximum and Minimum Crown Pressure. It is evident that as the pole distance (*i.e.* pressure at the crown) (Fig. 53) increases, the rays and corresponding strings become more nearly horizontal, and consequently the funicular polygon becomes more nearly flat. Hence of all lines of pressure which can be drawn within the middle third of the arch ring, that one corresponds to a minimum crown pressure which touches the outer curve at or near the crown and the inner curve at some point farther from the crown, while the line of pressure corresponding to a maximum crown pressure will touch the inner curve at or near the crown and the outer curve at some point farther from the crown.

86. Position of True Line of Pressure. An infinite number of funicular polygons may be constructed for a given system of arch

loads.   In general it is uncertain which of these is the true line of
pressure.   Of the various theories (see references) which have been
advanced to serve as a basis for determining the position of the true
line of pressure, the theory of *least crown thrust* seems to be most
commonly employed.   It is, therefore, briefly presented here.

This theory is essentially that the true line of pressure is that
one which, lying within the middle third of the arch ring, corresponds to a minimum crown pressure.   It appears to be based
upon the observation that most arches settle at the crown when the
centre is removed, and upon the assumption that the crown pressure
is a passive force developed by the tendency of the two half-rings to
tip towards each other, and is the least force that is necessary to
prevent such overturning.

If the arch settles at the crown (as a result of rotation, not sliding, of the arch stones), the obvious tendency will be to open the
joints as shown in Fig. 54, the resultant pressure moving upward at

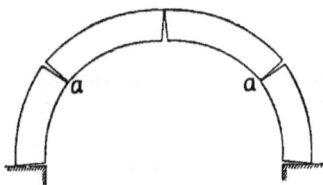

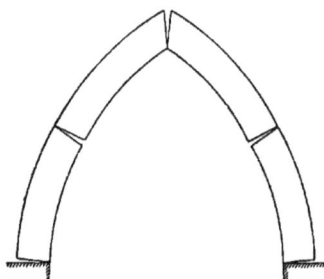

FIG. 54.   FIG. 55.

the crown and inward at the haunches.   This position of the line of
pressure corresponds to minimum crown thrust (Art. 85).   The
joints *a, a*, where the tendency to open at the extrados is greatest,
are called the *joints of rupture*.   They correspond to the points where
the line of pressure touches the inner limiting curve (Fig. 53).   In
the case of a full centred arch or elliptical arch, the joints of rupture
are approximately 30° from the horizontal.   In a segmental arch
subtending less than 120°, they are at or near the springing planes.

In the case of a pointed arch, or an arch very lightly loaded at
the crown and heavily loaded at the haunches, the tendency may be
for the crown to rise and the haunches to move inward (Fig. 55).

**87. Construction of True Line of Pressure according to Theory of
Least Crown Thrust.**   In Fig. 53 $XY'Y''$ is the true line of pressure
according to this theory, Y' being at the joint of rupture.   The

trial line of pressure may be drawn at once through X and Y', the joint of rupture being located as explained in Art. 86. If the line of thrust thus drawn does not fall within the middle third, the construction can be repeated as explained in Art. 84.

88. Example. Figure 1, Plate IV, is one-half of a symmetrical full centred arch in a masonry wall whose height is limited by a horizontal line, as shown. It is required to draw the line of pressure according to the theory of least crown thrust, and to determine if the arch satisfies the conditions of stability.

The half-ring is divided by radial lines, which need not coincide with the actual joints of the arch. It will be assumed that each of these divisions or voussoirs supports the weight of the portion of wall directly above it, as indicated by the vertical lines. If the specific gravity of the material above the arch ring is the same as that of the arch ring, the load supported by any voussoir, as mn, is proportional to the area of the polygon $mnn'm'$. If these specific gravities are unequal, the vertical ordinates may be altered in length so that the areas above the voussoirs will represent weights to the same scale as the areas of the voussoirs themselves. Otherwise the weights of the voussoirs and material above them may be dealt with separately. In this example, the wall is of uniform thickness and the weight of the masonry is, throughout, 160 lbs. per cubic foot. ·

Considering one foot thickness of wall, the loads (see table) are calculated by multiplying the corresponding areas by 160 lbs. HI is taken to include two voussoirs to avoid confusing the drawing; IJ is the·weight of the masonry to the right of the line i; JK is the weight of the masonry below RS.

The resultant loads act at the centre of gravity of the areas. The centre of gravity of the voussoir mn is O', and the centre of gravity of the trapezoid above this voussoir is O'' (see Art. 75). The centre of gravity O of the entire area $mnn'm'$ is then found by dividing the line $O'O''$ into parts inversely proportional to these areas. The centres of gravity are indicated by circles.

TABLE OF LOADS.
(FIG. 1, PLATE IV.)

VOUSSOIR.	WT. OF VOUSSOIR.	WT. ABOVE VOUSSOIR.
	lbs.	lbs.
AB	540	1360
BC	540	1440
CD	540	1570
DE	540	1700
EF	540	1790
FG	540	1760
GH	540	1540
HI	1080	1600
IJ	—	10240
JK	—	7680

The line of pressure is now constructed as follows: The loads
AB, BC, etc., are plotted to scale, and, selecting any pole P on a horizontal line through A, a trial polygon xy is drawn, the point x being
one-third the depth of the joint below the extrados. In drawing this
polygon, the intersections c', d', etc., of its various strings with the
string a are marked. These points of intersection locate the resultant load lying between the two intersecting strings; hence, the
strings of any other polygon for the given loads will pass respectively
through these same points, if the first string a is unchanged. The
line of pressure desired is such that the resultant pressure at the
joint of rupture will act at one-third the depth of the joint from the
intrados. The joint of rupture may be determined by trial, as
follows: Knowing that it is about 60° from the crown, we trisect the
joint g at 1, and draw the string g through 1 and g'. The adjacent
strings are then drawn, and it is thus found that f is the true joint
of rupture. This joint is then trisected at 2, and the string $2f'$ is
drawn. The polygon is completed by drawing the remaining strings
in succession through e', d', etc. The pole P' of this polygon is
located by drawing the ray FP' parallel to the string f. The
points of application of the resultant pressure at the different
joints are indicated by arrows. This line of pressure falls outside
the middle third of the joint, at the springing plane and first joint
above. This portion of the arch can be treated as part of the
abutment.

Aside from the condition that the pressure must act within the
middle third of the arch ring, the resistance to sliding and crushing
must be investigated.

As regards sliding, it is seen that the direction of the resultant
pressure at each joint is very nearly normal to the joint, with the
exception of the springing plane, where the pressure $p'i$ makes
with the normal an angle greater than $\tan^{-1}.4$. When, however, the
weight ij is combined with pi, the resultant pressure, $p''j$, satisfies
the requirement for safety against sliding.

For crushing, the maximum compression stress is to be calculated
at each dangerous joint. For example, the resultant stress on the
joint f, found by scaling off the ray $P'F$, is 13100 lbs. The area
of the surface of the joint is $1 \times 1\frac{3}{4} = 1\frac{3}{4}$ sq. ft. $= 252$ sq. in. Aver-
age pressure per sq. in. $= \frac{13100}{252} = 52$ lbs. Hence the maximum
stress $= 2 \times 52 = 104$ lbs. per sq. in. (See Art. 80.) This stress
must not exceed the working compression strength of the masonry.
(For strength of masonry, see Lanza's *Applied Mechanics* and other
references.)

89. Unsymmetrical Cases. When the arch or loading is unsymmetrical, the line of pressure is also unsymmetrical, and must therefore be drawn for the whole arch. The construction of the line of pressure involves the problem of drawing a funicular polygon through three points. (See Arts. 23 and 24.)

EXAMPLE. (Fig. 2, Plate IV.) Given a segmental arch of 16 ft. span and 3 ft. rise. Thickness of arch ring $= 1\frac{1}{4}$ ft.. The left and right halves are loaded with 3200 lbs. and 6400 lbs. respectively, these loads being uniformly distributed over the arch ring. It is required to determine the possibility of drawing a line of pressure within the middle third of the arch ring.

The arch is divided into 16 equal voussoirs, and the load supported by each is assumed to act at the middle of its outer surface. The loads are plotted to scale, and the funicular polygon xy is constructed, using P for pole. Selecting the points t', v', and u' at one-third and two-thirds the depth of the joints from the intrados, the funicular polygon which will pass through these points is located. This polygon is not drawn, but the points on it, falling outside the middle third, which serve to locate the final line of pressure, are marked by circles. These points were determined by the method of Art. 24. It is seen that, to the right of the crown, this polygon rises above the middle third, while near the left abutment it falls below. From the position of these points it appears that a polygon drawn through the three points $1'$, $2'$, and v' will probably fall within the specified limits. The pole P' of this final polygon was located by the method of Art. 23. PZ and PZ' are drawn parallel to 1, 2 and 2, v respectively; then ZP' and $Z'P'$ are drawn parallel to $1'$, $2'$ and $2'$, v' respectively. The point of intersection of these lines is P'. From P' as pole the polygon is constructed so that the strings e, k, and s pass respectively through the points $1'$, $2'$, and v'. This polygon lies wholly within the middle third of the arch ring.

The location of this polygon corresponds to that of the true line of pressure of a symmetrical arch according to the theory of least crown thrust, with the exception that it does not touch the outer limiting curve at the crown. Safety as regards sliding and crushing is investigated as previously explained.

90. General Remarks. (See also Lanza's *Applied Mechanics*, § 270.) The investigation of the stability of a voussoir arch is necessarily inexact.

(1) The loads are more or less uncertain, both in amount and

distribution. In the example (Fig. 1, Plate IV) it was assumed that each voussoir supported the weight of the entire mass above it. This is a common assumption in dealing with bridge arches. In some cases such an assumption would be manifestly absurd; *e.g.* an arch in a high masonry wall, or a tunnel arch deep under ground.

(2) The location of the true line of pressure is uncertain. The method already explained for locating the line of pressure is conventional. It is not derived mathematically. It has been proposed by different writers to place the determination of the line of pressure upon a rational basis, by treating the masonry arch as an elastic arch fixed at the ends. It is not fully evident that the voussoir arch conforms sufficiently closely to the case of an elastic arch with fixed ends to make such treatment reliable, even assuming that the loads are accurately known. (For discussions of arch theories, see References.)

The stability of the abutments is essential to that of the arch, and must be considered in connection with the arch.

§ 3. *Abutments, Piers, etc.*

91. Conditions of Stability. The general conditions of stability of § 1 are applicable to any piles of masonry subjected to the action of external forces, such as the thrust of an arch or truss, pressure of earth, water, wind, etc.

92. Example 1. Fig. 56 is an abutment subjected to a horizontal pressure *AB* and a vertical pressure *BC*, their resultant, *AC*, acting at the point *C*. *CD, DE, EF*, and *FG* are the weights of the

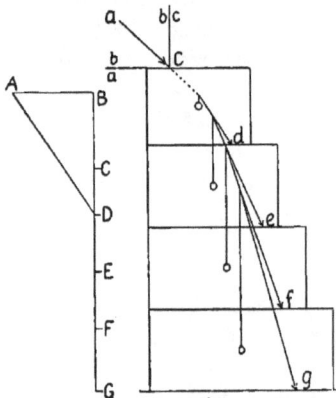

FIG. 56.

four divisions of the masonry, their centres of gravity being marked by circles. The pressure on the joint *d* is the resultant of *AC* and the weight *CD* of the first block. This resultant is *AD*, its line of action passing through the intersection of *ac* and a vertical through the centre of gravity of the block. The point of application of this resultant pressure, at *d*, is indicated by an arrow. The resultant pressures at the remaining joints are found in a similar manner.

The conditions of safety as regards sliding, overturning, and crushing have been previously discussed. The maximum pressure on the soil must also be kept within safe limits. (See Baker's *Masonry Construction*, Chap. X.)

A broken line connecting the points of application c, d, e, f, and g of the resultant pressures on the successive joints of an abutment is called the *line of resistance* or *line of pressure*, as in case of an arch. (See Art. 82.)

93. Example 2. The abutments of a masonry arch can be considered in connection with the arch. In Fig. 1, Plate IV, the weights IJ and JK, lying above and below the joint RS, are plotted to half scale (4000 lbs. $= 1$ in.), the pole for these two loads being located by bisecting the ray $P'I$ at P''. The resultant pressure on RS is $P''J$, its line of action $p''j$ passing through the intersection of $p'i$ and ij. The resultant pressure on the base is $P''K$, its line of action being $p''k$. To find the maximum intensity of the pressure on the base we have $P''K = 9.1$ (inches) $\times 4000 = 36400$ lbs. By measurement, $p''k$ acts 1.1 ft. from the centre of the base. The bearing area is $7\frac{3}{4} \times 1 = 7\frac{3}{4}$ sq. ft. Substituting these values in the formula $p = \dfrac{P}{A} + \dfrac{Px_0 a}{I}$ (Lanza, § 207), the maximum intensity of the pressure is found to be 60 lbs. per sq. in.

94. Example 3. Let Fig. 57 represent a pier supporting the thrust of an arch on each side. These thrusts are AB and BC, their point of intersection O lying on the centre line of the pier. Let $CD, DE, EF,$ and FG represent the weights of the pier divisions, CD being the weight of the pier masonry above the joint d. The resultant pressure on d is AD, its line of action Od passing through O. The lines of action of the pressures on the remaining joints will also pass through O, since the vertical through this point contains the centres of gravity of all the blocks. These lines are $Od, Oe, Of,$ and Og, drawn parallel respectively to $AD, AE, AF,$ and AG. If

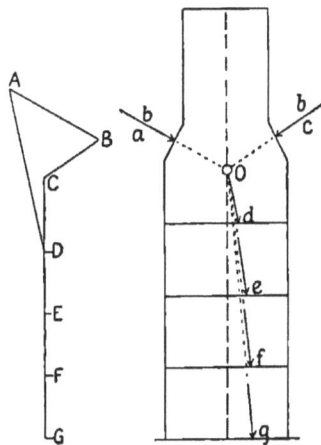

Fig. 57.

the thrusts of the two arches are equal and equally inclined, the

resultant pressure on the pier will evidently be vertical and equal
to the weight of the two half-arches besides the weight of the pier
itself.

95. Example 4. Let Fig. 58 represent a chimney subjected to
wind pressure. The weights of the portions ab, bc, etc., are AB,
BC, etc., and the wind pressures on these portions are AB', $B'C'$,
etc. The lines of action of these wind pressures are the horizontal

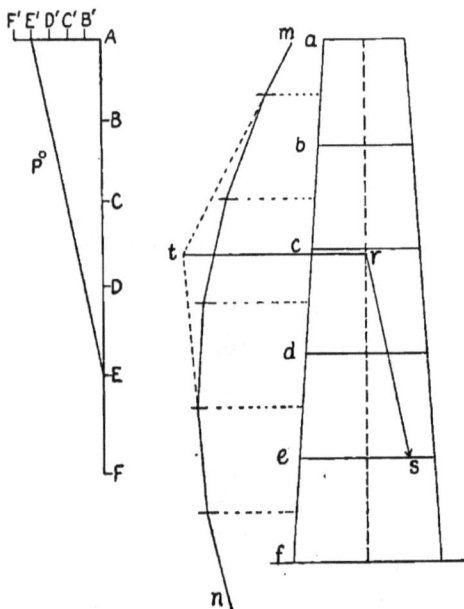

FIG. 58.

lines ab, bc, etc. With any pole P draw the funicular polygon mn.
(For this purpose the wind pressures should be plotted to a larger
scale.)

To find the resultant pressure at any section, as e, the line of
action tr of the resultant wind pressure above that section is located
by the intersection of the strings a and e.

The pressure at the section e is the resultant of this wind pressure and the weight of masonry above e. The line of action rs
of this pressure will act through r, the point of intersection of the
resultant wind pressure and weight of masonry above e, and its
direction will be parallel to $E'E$, its magnitude being represented by

the length of $E'E$. The resultant pressure at any other section can be determined in a similar manner.

The preceding examples will serve to indicate the method of determining the stability of such structures when the loads are known.

REFERENCES.

The following list includes elementary works in which more detailed explanations will be found, and others of a more advanced character. The list is not intended to be exhaustive.

"Graphic Statics," by Mansfield Merriman. John Wiley & Sons, New York.

"Elements of Graphic Statics," by L. M. Hoskins. Macmillan & Co., New York.

"Construction of Trussed Roofs," by N. C. Ricker. W. T. Comstock, New York.

"Statique Graphique," by Rouché. Baudry & Co., Paris.

"Charpentes Metalliques," by Dechamps. Vaillant-Carmanne, Liège.

"Theory and Practice of Modern Framed Structures," by J. B. Johnson. John Wiley & Sons, New York.

"Applications de la Statique Graphique," by Koechlin. Baudry & Co., Paris.

"Treatise on Masonry Construction," by I. O. Baker. John Wiley & Sons, New York.

For additional illustrations of roof construction, see

"Revue Technique de L'Exposition de Chicago," Part I, Architecture (with Atlas). E. Bernard & Co., Paris.

Publications of Iron Construction Companies.

PLATE I.

Fig. 2. (See p.   ).

Fig. 1 (See p.   ).

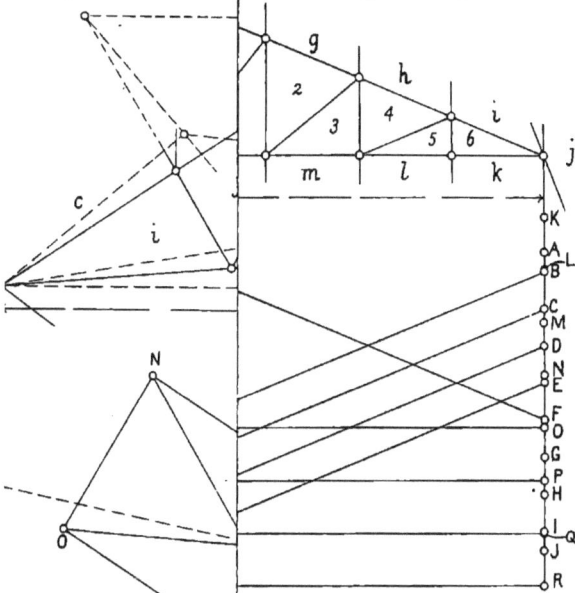

Load.

Fig. 2b. Wind Roller Sid

Scales. Fig
5 ft. = 1 inc
0   5

3000 lbs. = 1 in
0   3000

Fig. 1b. Snow Load.

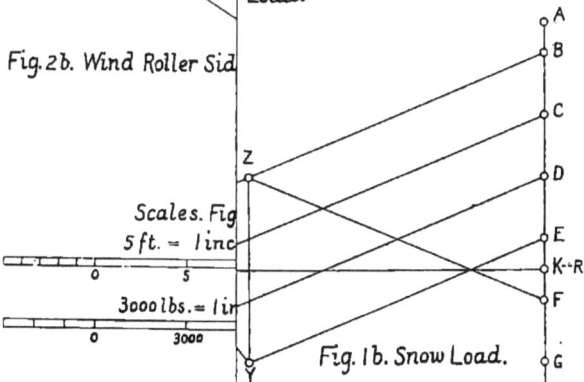

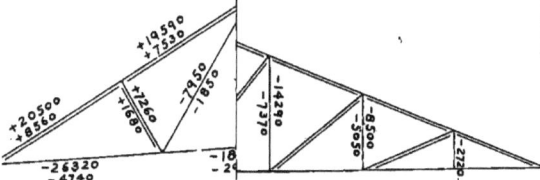

Fig. 2d n. Stresses.

PLATE I.

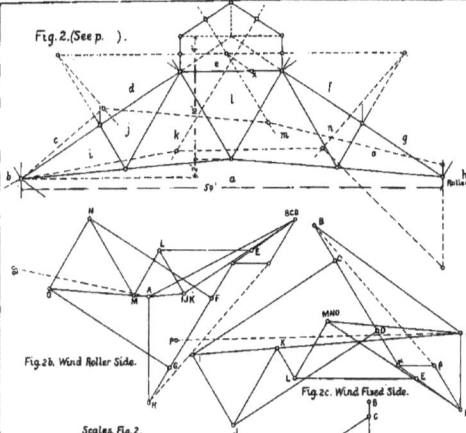

Fig. 2.(See p.   ).

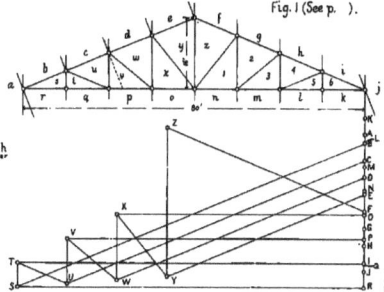

Fig. I (See p.   ).

Fig. 2b. Wind Roller Side.

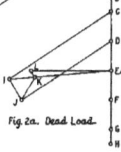

Fig. 2c. Wind Fixed Side.

Scales. Fig. I.
10 ft. = 1 inch.
5000 lbs. = 1 inch.

Fig. 1a. Dead Load.

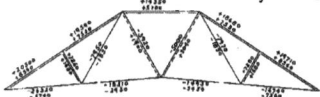

Scales. Fig. 2.
3 ft. = 1 inch.
5000 lbs. = 1 inch.

Fig. 2a. Dead Load.

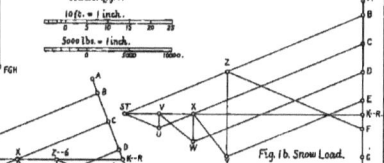

Fig. 1b. Snow Load.

Fig. 1c. Wind Left.

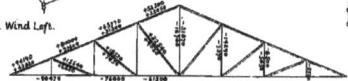

Fig. 2d. Max. and Min. Stresses.

Fig. 1d. Max. and Min. Stresses.

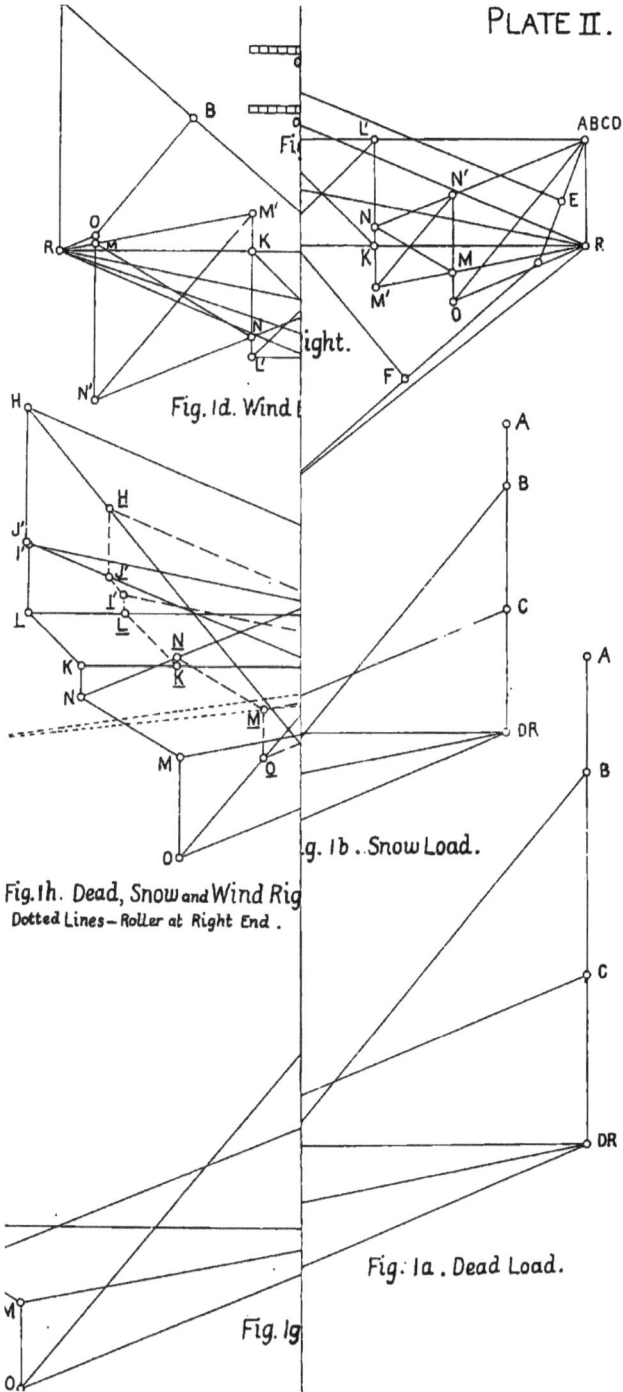

PLATE II.

Fig. Id. Wind

Fig. 1b. Snow Load.

Fig. 1h. Dead, Snow and Wind Rig
Dotted Lines — Roller at Right End.

Fig. 1a. Dead Load.

Fig. 1g

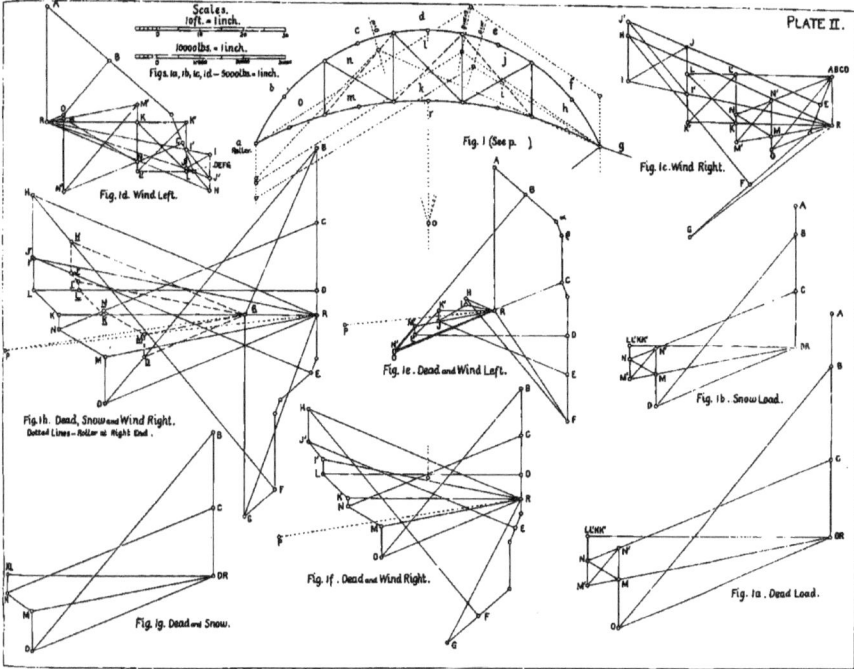

PLATE II.

Scales.
10ft. = 1inch.

10000 lbs. = 1inch.

Figs. 1a, 1b, 1c, 1d = 5000lbs.= 1inch.

Fig. I (See p.   )

Fig. 1c. Wind Right.

Fig. 1d. Wind Left.

Fig. 1e. Dead and Wind Left.

Fig. 1b. Snow Load.

Fig. 1h. Dead, Snow and Wind Right.
Dotted Lines = Roller at Right End.

Fig. 1f. Dead and Wind Right.

Fig. 1a. Dead Load.

Fig. 1g. Dead and Snow.

PLATE III.

Fig. 1

4000.

b

I = 100   I = 150   20

(1)

Fig. 1a'

C
B
A

a'   b'   C'   d'

Fig. 1b

(I')

C'

n'   o'   b'

r''

S'

t'   a'   z'

u'

V'

w'

Fig. 2

a   b   c   d   e   f

l

a'

b'

c'

d'   e'

P'

Fig.

N

S   T   R   O   M

K   L

W   V

I

G
G'   H

F'

J

F

E'

A
B
C
L
D
E
F
G
H
I
J
K

Fig. 2a'

P

P'   D'

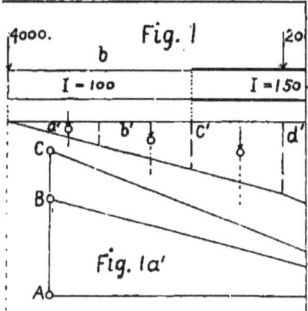

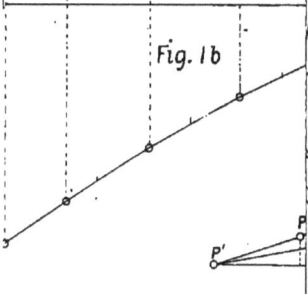

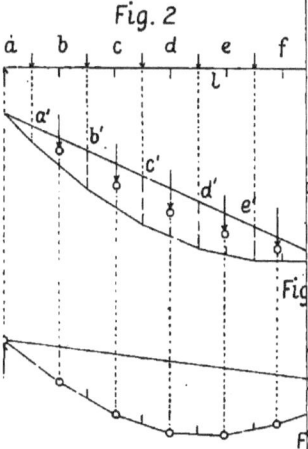

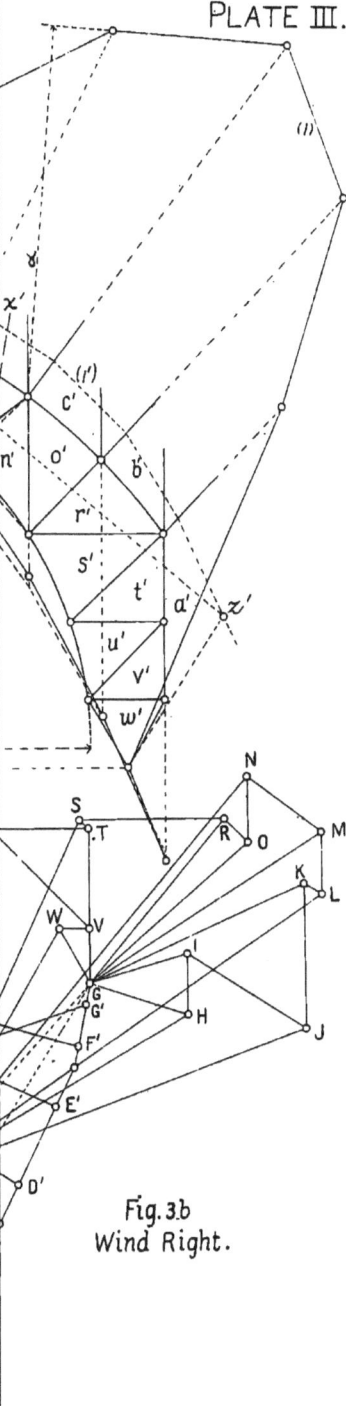

Fig. 3b
Wind Right.

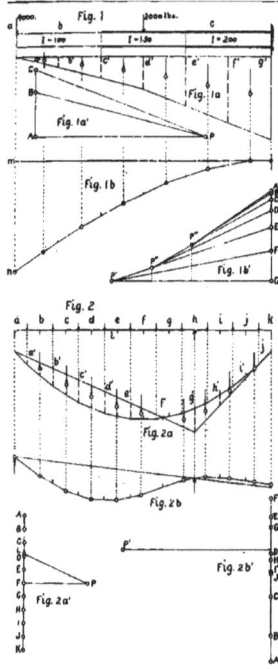

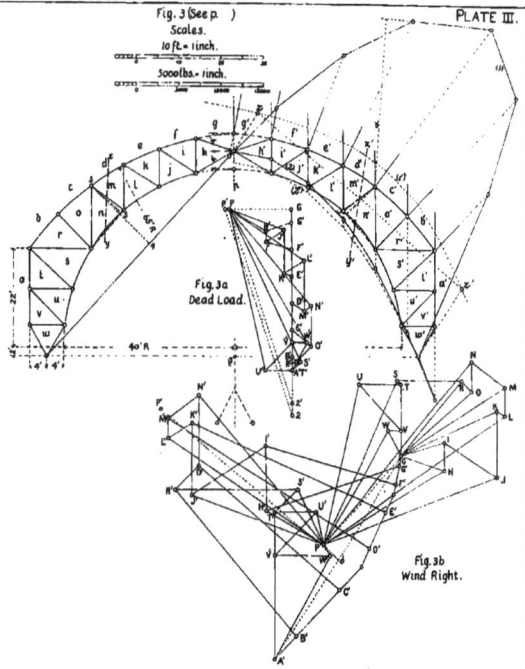

PLATE III.

PLATE IV.

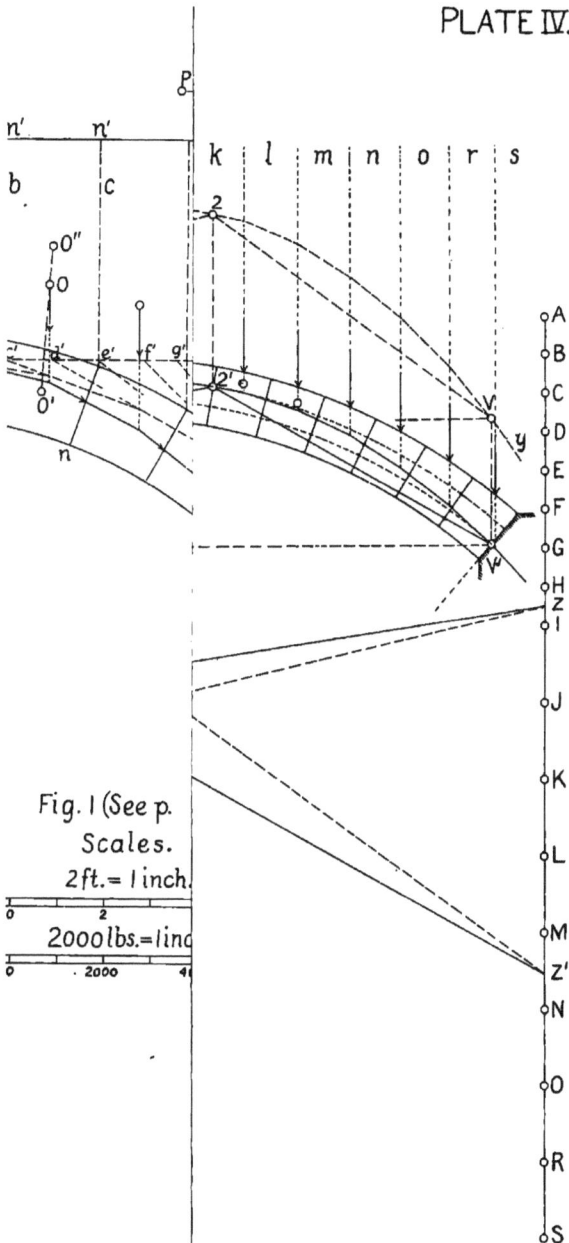

Fig. I (See p.
Scales.
2 ft.= 1 inch.
2000 lbs.= 1 inc

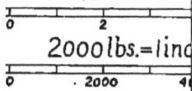

PLATE IV.

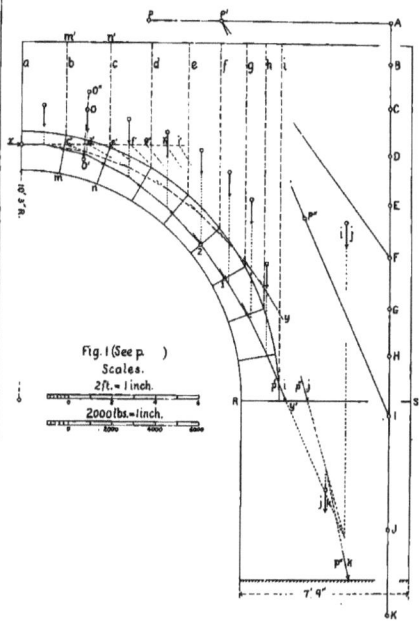

Fig. I (See p.   )
Scales.
2 ft. = 1 inch.
2000 lbs. = 1 inch.

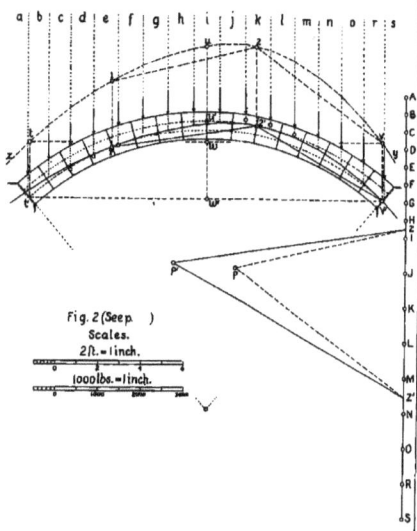

Fig. 2 (See p.   )
Scales.
2 ft. = 1 inch.
1000 lbs. = 1 inch.